THE JOY OF DISCOVERY

Great Encounters Along the Way

THE JOY OF DISCOVERY

Great Encounters Along the Way

Walter Thirring

University of Vienna, Austria & CERN, Switzerland

World Scientific

NEW JERSEY • LONDON • SINGAPORE • BEIJING • SHANGHAI • HONG KONG • TAIPEI • CHENNAI

Published by

World Scientific Publishing Co. Pte. Ltd.
5 Toh Tuck Link, Singapore 596224
USA office: 27 Warren Street, Suite 401-402, Hackensack, NJ 07601
UK office: 57 Shelton Street, Covent Garden, London WC2H 9HE

British Library Cataloguing-in-Publication Data
A catalogue record for this book is available from the British Library.

This English edition is a translated version of the German edition *Lust am Forschen. Lebensweg und Begegnungen.*

THE JOY OF DISCOVERY
Great Encounters along the Way

ISBN-13 978-981-4322-96-6
ISBN-10 981-4322-96-2

Typeset by Stallion Press
Email: enquiries@stallionpress.com

Printed in Singapore by Mainland Press Pte Ltd.

Dedicated to
my sons Klaus and Peter

Contents

The days of your years are threescore years and ten;
and if by reason of strength they be fourscore years,
yet is their strength labor and sorrow;
for it is soon cut off,
and we fly away.

(*Psalm 90*)

Preface

The 20th century was both a horrible as well as wonderful time. After the insanity of the Nazi dictatorship, the voice of reason prevailed and presented Central Europe with the longest period of peace in its history. As soon as the human spirit has the opportunity to grow and develop, it will begin to notice the secrets of the world we live in. It will hone its perception so that it can break down matter in an interplay of exactly balanced parts and let the firmament become a horde of escaping galaxies. It uses all of this knowledge to improve itself and its abilities until they reach immeasurable heights. And this not only in the material sense where everything is constantly getting bigger and faster, but also with regard to intellectual questions. The human spirit developed ways to conduct calculations in fractions of a second which would ordinarily take decades, and can store the information found in entire libraries on chips the size of a trouser button. It is well on its way to deciphering the secrets of life itself, and before we know it the species *homo sapiens* will have undergone a metamorphosis.

As I have had the privilege not only to experience this turning point firsthand, but also to get to know personally the main protagonists of this development from Einstein to Hawking, I felt I had to tell this story of successful human endeavor.

Of course, my memoir contains only my own point of view, or rather snippets of thought that have gathered over time. I hope that the reader will be able to find something of interest or amusement in this more or less random selection. Where possible, I have omitted things like romances, illnesses and honors. If not for science, however, a central driving force would be missing. I have put scientific explanations in separate boxes so that the text is not bogged down with them. This way, those who are interested can find out more about what I'm talking about, while those less ambitious readers can simply ignore these parts and continue reading.

I am not a professional writer and many people have helped me in writing this book. On all of their behalf I would like to express my deepest thanks to Dr. Lore Kummer, Dr. Helga Engin-Denitz, Dr. Gerlinde Faustmann and Dr. Josef Streicher for their patient proofreading as well as to Dr. Brigitte Buschbeck for the beautiful photograph of particle reactions. Dr. Wolfgang Reiter gave me valuable advice for the overall concept of the book.

Family Background/Childhood (1927–1938)

Fig. 1: The author today and as a child

While I was rummaging through a desk drawer, I came across an old cardboard box. I thought it would be filled with junk waiting to be tossed out, nevertheless, my curiosity was aroused, and I took a look inside. I found some faded yellowed photographs. One showed a small tyke all bundled up with a stick in his hand, possibly to chase away pigeons? It might be me, although it's hard to see the similarity. The photograph must have been taken in a park... but now the park looks completely different. How everything has changed — Vienna and the people in it, the whole world. The little boy had no way of knowing what this restless, troubled world would hold in store for him.

When I look back on that time, I don't only see myself chasing away the pigeons. Another memory comes to the fore. Here I am, walking with my mother like a good boy going to play in the Arne Carleson Park. Suddenly, a man dressed in rags appears before us. Without any warning he collapses to the ground and can't get up anymore. My mother rushes over and was able to gather from his babbling that he hadn't had anything to eat for two days and couldn't go on anymore. Now, because I was a bit frail, I always had to carry around a mid-morning snack, and I was always glad when I didn't have to eat it. My mother gave my bread to the man which revived him enough so that he was able to sit down on a bench. I don't know what happened to him, but this encounter taught me that during the international economic depression in 1932, it was not a given that everyone had bread to eat. The depression and its aftermath, as well as the civil war in 1934, took place beyond my childish horizons. I heard the shooting in the streets and wasn't allowed outside for three days, but what it was all good for, why people were suddenly allowed to shoot at each other, that, of course, I couldn't understand (and I still don't understand it today). I first heard that all was not well in our world in 1937, when the great physicist Max von Laue came to see my father. While they were on a hike he told my father about the horrible direction politics had taken in Germany and what a dangerous man Hitler was.

Nevertheless, I remember feeling secure and sheltered when I think of my childhood. I wasn't fully aware of the aberrations on the political stage; my world was composed of my family, and it was a cozy, perfect world. I was only seeing a snapshot, though, and couldn't guess all of the sweat and hard work that had accompanied their long way up or the tears shed at the sudden collapse of the Hungarian line. In any case, let me begin by clarifying my roots:

The Thirrings

Fig. 2: The Thirring rocks near Dobogo-kö

For the common people, the history of Central Europe is one of the horrors of wars, conquests and banishments. The first documented Thirrings were likewise swept into the Danube monarchy as a consequence of the Thirty Years' War, which destroyed everything in its path. The war began heading north, and the master shoemaker Mathias found that life in Thuringia (Thüringen) was becoming too dangerous. He sought refuge in the south. However, he was not allowed to stay in Austria, as he had the wrong religion. So he moved on, and in 1623 he received the rights of citizenship in Sopron (which was known as Ödenburg at that time). Less than an hour's drive away from Vienna today, it was then a different world, a place of tolerance for the Protestants from the north; here, you didn't have to be Catholic. The only problem was caused by the family name, the Hungarian church scribes didn't quite know what to make of it and on the books you'll find Thüring, Thiring, Diring, Düring, Türinger, Thüringer, Düringer, Diringer, until the spelling Thirring prevailed.

The family prospered and the Thirrings soon began to work as tailors and weavers and all kinds of craftsmen. They even were able to achieve a certain level of wealth, and were sponsors of the 2,000 seat Protestant Church in Sopron. They eventually owned a patrician villa near the moat and the leading hardware business in Sopron, which allowed them to finance the intellectual pursuits of the later generations.

The Hungarians were able to profit from letting the Protestants stay, because they made rich contributions to the general level of learning and education. Many Protestant schools were founded which later developed into elite schools. For example, prominent scientists Johann von Neumann and Eugen Wigner were products of the Protestant secondary school in Budapest.

The Protestant lyceum in Sopron changed the character of the Thirring family. The family of craftsmen became a family of intellectuals. The prominence of the family was no longer limited to one place, but spread throughout the country, and through its scientific influence, ultimately throughout the world.

The first to rise above the provincial level was Gustav Thirring, a cousin of my grandfather. After graduating from the Protestant lyceum in Sopron, he went to university in Budapest and climbed the academic ladder of PhD, lecturer and Professor Extraordianrius. He ultimately became the director of the Hungarian Central Bureau of Statistics. Thanks to his versatility, he also became the president of the Hungarian Alpine Association. To the west of Budapest lies the counterpart to the Vienna Woods, just a little farther away and about a thousand feet higher. Close to the highway in Dobogokö there's an inn, where above the entrance there's a gentleman with a stand-up collar and a straight mustache confidently looking down and greeting the guests, that's Gustav Thirring. Going further into the woods, two mighty rock formations suddenly appear and these are known as the Thirring Rocks. He was a renowned figure, so that when I give a lecture in Budapest, I'm not introduced as Walter Thirring, but as the great-nephew of Gustav.

Julius Thirring, my grandfather, was also a graduate of the Protestant lyceum in Sopron and went on to study mathematics in Vienna and Berlin, completing his studies under Charles Hermite at the Sorbonne. However, that was also the end of his academic career. He had fallen in love in Vienna with the daughter of the court photographer of Archduke Maximilian, the future Emperor of Mexico, and had to take a more lucrative position. He became the principal of a secondary school and led a quiet life. He wasn't a man of large stature, but he commanded an atmosphere of discipline and order. I always remember him with a stand-up collar and elaborately twirled mustache. He was also multi-talented — he translated Hungarian literature into German and could have become a great violinist had he not been so athletic. He was able to jump higher than his height (which, to be fair, was not that great) and was considered to be a pioneer in riding penny farthings. This was a relatively dangerous thing to do and he completely shattered his left index finger in a fall. That put an end to his career as a violinist and he had to be satisfied with a reputation as an excellent chess player. His perfection of the Hartlaub gambit can still be found in the international chess literature and a game he played against world champion Carl Schlechter in 1893 is considered a classic example on the Internet. He had four children. His second son died of diphtheria as a child, but the others turned to the natural sciences. Hans became a physicist, Ernst and Grete became biologists.

This is a rough sketch of the rise of the Thirring family over the long period of three centuries. The fall of the Thirrings in Sopron went much faster. If they had not been exiled to Austria, the line would have completely disappeared. The Hungarians' tolerance had slowly started to crumble. In 1948, the communists came to power and the Thirrings were once again in the wrong boat — this time in the wrong class. The hardware business was taken over by the state and its owners made lesser employees. During this time some changed their name to the more Hungarian Tass, to at least be rid of the wrong name. Many others fled all over the world. When I visited the Thirring villa the last time I was in Sopron, none of the younger people I spoke to had ever heard of the name Thirring. Just one old woman told me that the last Thirring had moved out years ago.

My Mother

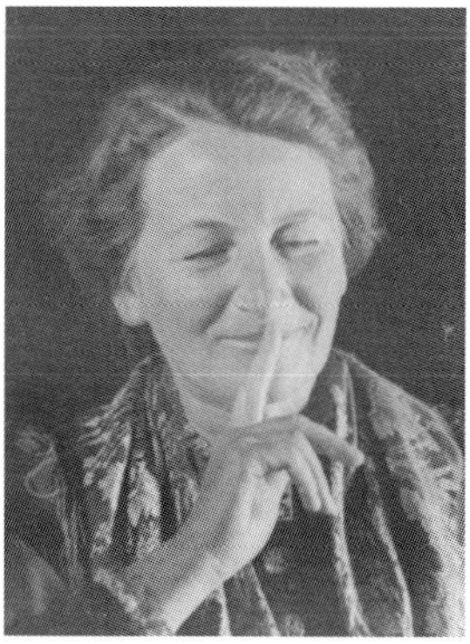

Fig. 3: Antonia Thirring

My parents belonged to that rare species of the true Viennese who were born in Vienna. My mother's father was from Bohemia and was called Kriz (or in German: Krisch), her mother's name was Horak, which sounds Bohemian, although it still qualifies as Viennese. My other grandmother was from Trieste, which makes you think she'd be Italian. That wasn't the case; her name was Malovich, which sounds more like it's Croatian. In any case, that is the raw material from which the metropolitan melting pot created the Viennese.

Old photographs show grandfather Krisch as a smart-looking bookkeeper. From this job he went on to have a career as a managing accountant for a furniture store, which was so lucrative that he could afford to raise eight children. The burden was mainly borne by my grandmother. I can't recall her having any help. On the other hand, for the standing of my grandparents on my father's side it was a given that they had a live-in maid. However, at least all three Krisch boys were able to go to college and found high level positions accordingly. In any case, eight children are quite a lot and at times the number of beds in the apartment was lower than the number of occupants. Their small front garden was an El Dorado for the children, and much later on his orchid plantations on Hawaii, my uncle was able to say that he had been raising orchids for half a century. My mother also studied, she was trained to be a cooking instructor. This came in handy during the great hunger after the First World War, and she became the director of a public kitchen. Many university employees also frequented the kitchen. My father liked it so much that he remained a loyal customer and thus my parents got married in 1921. After the collapse of the monarchy, civil servants were, of course, not paid well, but professors still received much more than those in less qualified positions. My mother left her job and devoted herself entirely to her family. For my brother and me she was a loving, not a domineering mother. My brother unfortunately did not return from the war. I myself left the country when I was 22, so that she was the only one there for my father. She was everything for him: housekeeper, writer, reader, traveling companion and a constant, loyal, tolerant wife who only rarely complained and put on a brave face to bad times. However, making dumplings all day long did not fulfill my mother intellectually. Every day she spent several hours reading German or English classical literature. My library has her to thank for its wonderful old classics editions. As the daughter of an educated household, she naturally learned how to play the piano. Her talent; however, far exceeded the ordinary standard; her teacher was a student of Anton Bruckner and she had adopted his stance that Bach's Inventions should form the basic training of every pianist. I often heard her play these as a child and they formed my understanding of music. When I later studied the Inventions, I found that I already knew them by heart.

My Brother

Fig. 4: Harald Thirring

My brother, three years older than I, was the genius in the family. He proved to be highly gifted in school. His interests were broad, stretching across the sciences and the liberal arts. For my tastes they were even too broad. For example, he enjoyed playing with tin soldiers, which I found mindlessly dull, but any comment I made in this direction threw him immediately into a temper tantrum. Otherwise, we had a close relationship; I had no problem acknowledging his superiority. Naturally, this just drove me to emulate him, but as soon as I had begun to approach his intellectual level, he had to go into the army and was lost to me. What I have left from him is a thick folder of APO letters. I'm still amazed today by the maturity in his style; after all, he was still a teenager. As a sample of this I've written down verbatim his farewell letter written while he was in the static warfare at the Eastern Front.

Dear Parents! *February 4, 1944*

I don't think there has ever been a harder letter for me to write than this one, which should be a farewell letter to you, whom I love so much. Now that I'm writing it, I still think it's impossible, but a dark voice is warning me to think of every eventuality.

Don't be so sad, I'm now in a better world without machine guns and canons, without the GPU[a] *and the Gestapo, without war and hate. And remember that I might not have to go to Siberia or to some Soviet prison. And believe me that my life has been so wonderful until now, that at least it won't be clouded by the hard times after the war. It's true that all of my big plans will come to naught, but you still have Walter, and he has just as much talent and creative will combined. He'll make it, he is the future.*

Yes, I have one last request; my estate is so negligible that it would be ridiculous for me to start listing particulars; everything in its entirety should go to Walter. So goodbye everyone and think about me from time to time.

Best wishes for all of you,

Your Harald

A few days after he had written this letter he was hit by a bullet. He survived, and if Count Stauffenberg had been successful on July 20th, or if the Americans hadn't dithered for two years with building the atomic bomb, then his dark voice would have been wrong. However, this was unfortunately not to be and, with the exception of a short message in January 1945, all traces of Harald Thirring disappear. This was when the Russians had pushed through to the Oder river and the German army was throwing everything it had at them. My brother was shipped out there on overnight expresses. He had been trained as a military engineer, a sapper, but had to hold the command as an infantryman at the part of the western bank of the Oder river where the Russians would build a bridgehead. There was no way out for my brother. The Russians were in a position of overwhelming superiority; retreat was

[a] The GPU was the Soviet secret police, the predecessor of the KGB.

high treason, and surrender useless, because when the Russians were fighting for a bridgehead they would not want to take any prisoners.

Just one member of his unit was heard from again. He had managed to escape into the bushes. Since then there have only been rumors about Harald, that he was seen in various POW camps, but never any word from Harald himself.

My Father

Fig. 5: Hans Thirring

My father considered himself to be a philosopher, a learned man, and wanted to share his knowledge with my brother and me even when we were children. Every night before we went to sleep he would lie down with us in bed for an hour or so of erudition. This was when he would explain all sorts of things to us. He had made up a saying to memorize the periodic table of the elements, and up to today his "Hahe liebe Bocenofne..." still comes in handy sometimes. He had discovered that the chemical symbols H, He, Li for the elements rhyme so well and are easier to remember this way.

He wasn't so happy with me in school as my erratic set of interests weren't always in line with his views. In elementary school I was more fascinated by building gliders than studying or practicing the piano. When my report card proved this, my father threatened not to let me go on to college, but said that I would have to be a cobbler. Since I was good with my hands, this occupation seemed to be more in keeping with my abilities, and I heartily agreed to his suggestion, which wasn't his intention at all.

Right after this, I became enthralled by music, wasting too much time on it in my father's opinion. He told me that most people in our family had been musical, but were all amateurs. I wasn't convinced by this argument at all; on the contrary, I thought it was high time that one of us began to be professional. My more gifted brother could carry on the physics tradition in the family. At this point my father gave up trying to influence me, but later it turned out differently anyway. My brother wrote to me from the battlefield that he would not be returning and it was up to me to carry on the scientific tradition.

My father bore the injustice inflicted upon him and so many others during the Nazi period with stoic aplomb. Never did he take out his anger on others or on the family. The affable manner he showed towards everyone came to him in good stead even with the regime. The Nazi party block warden often came to inspect our home, but my father was very quick at turning off the foreign radio station and sliding his papers in a desk drawer all the while engaging the informant in conversation and distracting him with his charm, so that he was never caught. He would have easily been able to sort out any political difficulties, as Hermann Göring was a boyhood friend of his, and Göring, of course, managed to arrange completely different kinds of things at this time as well.

But my father wanted to have nothing to do with this regime, including even an intervention from it. Our connection to the Göring family could be traced back to my grandfather's passion for penny farthings. While he was studying at the university in Berlin, he became friends with a medical student named Hermann Epstein. Epstein followed my grandfather back to Vienna and also became hooked on penny farthings. They decided to explore Austria on penny farthings. With the street conditions at that time, this sounds more like a nightmare, but they apparently particularly liked Mauterndorf and its castle. In any case, the future Baron Epstein whipped out his checkbook and bought the castle on the spot. He was able to as one of his ancestors was a chamberlain of Queen Luise. Ever since then, their family's wealth had continued to grow. Mauterndorf was not his only castle. But even this purchase didn't help because he suffered from depression. According to family legend, he was unhappily in love with my grandfather's sister. He decided to have a change of scenery and embarked on a trip around the world. In South West Africa he became friendly with the German Governor and his wife. This was Hermann Göring's father, although he was yet to be born. The Governor's wife was just pregnant with Hermann's oldest sister when the Görings were warned of a local insurrection. The African streets were getting too dangerous for them and they fled. They made it to Walvis Bay, where Hermann's sister was born. After they had managed to return to Germany, Epstein wanted to return their hospitality and invited them to stay at his castle in Mauterndorf. The castle was big enough to invite the Thirring family as well, and this was how the contact between the Görings and the Thirrings began. My father became friends with the Göring's younger daughter Paula. This friendship stood the test of time and even during the war Paula didn't shy away from seeing him, although he was persona non grata.

My father's optimism was unshakable. When the Americans reached Kitzbühel, where he was living at the time, towards the end of the war, he offered his services as an interpreter between the American military and the local authorities for free, since there was no pay. My father knew from the Voice of America that Austria was going to be treated as a liberated and not as a conquered country, but it seemed that the Americans didn't really know exactly the difference between where Austria was and where Germany was. This meant that they stole from the billeted houses just like other troops in enemy territory, and my father's job wasn't easy. Even when the American colonel screamed at him, calling him a war criminal because he repeated the request of the hunters to keep their guns, he didn't waver in his responsibility.

For my father, the USA represented the ideal liberal and democratic country. He liked to tell the story about how, as a young man, he traveled with the Vienna University choir to give a concert tour in the USA. This was before World War I. At that time there was no such thing as a US visa; you didn't even need a passport. You just needed to hand over an immunization certificate when entering the country, but my father had his buried deep in his suitcase. He put his hand in his pocket and found the menu from the last dinner on board. This was accepted as well, and he was able to enter without any problems. Later this would no longer be so easy for my father for the following reason. During the Nazi period the Vienna math department was "cleansed" of people like Helly, Tauber, Gödel and Menger and a certain Huber called the shots. He was a man of unknown scientific merit, but his politics were clear.

After the war, the professors fired by the Nazis were re-hired, and in 1946 my father became the Dean of the University of Vienna. It was incumbent upon him to eradicate what the Nazi period had spawned, and Huber was thrown out. However, Huber was quickly able to find a job with the CIA and denounced my father as a communist. Nothing could have been further from the truth; my father knew all about the crimes of Stalinism. During the war Fritz Houtermans came to see us. He came directly from the Soviet Union and was able to tell us all about the human rights situation in the workers' paradise. However, the Americans trusted Mr. Huber, and my father would have problems getting a US visa for a long time afterward. He remained tainted with the suspicion of having been a secret communist, and continued to be suspect in some circles. How else can it be explained that he never became a full member of the Austrian Academy of Sciences? It would seem that those people who were named between 1938 and 1945 and who could no longer be removed afterwards were able to prevent this in the secret ballots. In any case, he was too sure of himself to ever waste time talking about such things.

The picture I have of him from his later years is of him sitting at his desk, writing and writing and writing. He had to share his thoughts with everyone. Everyone was a welcome ear; he read from his manuscripts at lectures, in front of the family, to our shoemaker in Kitzbühel and I don't even know to who else. This was where his didactic talents came to shine. His lectures on physics were always a paradigm of intelligibility, and his audiences were inspired. His fight against the atomic bomb and his warnings of the energy crisis did not at first meet with results. He was written off as naïve, but he knew that his time would come.

My father had an entrepreneurial spirit. He was constantly inventing and improving things. He applied for countless patents: for "Seleenzellen", small electronic devices which were profitable because they provided the music for silent movies; for a rail tire, an electric shaver, an all-metal ski binding, a gas turbine — all of which were money pits — until he finally came up with something that caught the attention of the media, the Thirring coat "Flattermantel". The Thirring coat would blow up like a parachute from the airflow created while skiing. It wasn't intended to pull you up, but just to slow you down. It's attached down at the ankles, and it increases the air resistance of your body by up to 300%. Then the drop can't be steep enough. If a body were to reach the top speed of 125 mph (200 kph) in a vertical fall, with the Thirring coat this would be reduced by a factor of 3, so a little over 40 mph (65 kph), which is not terrifyingly fast. I used it to go straight down the northern slope of the Steinbergkogel in Kitzbühel, a pure pleasure. The coat did not turn out to be a big seller, though. At first it managed to become a fixture in movies and magazines, but then came the war and textiles were rationed so that it could no longer be produced. When the economy recovered after the war, skiing became popular among the masses. The slopes were soon so crowded that no one was able to do a straight downhill run anymore. Hence this invention also did not reach the popularity he'd hoped for.

Fig. 6: The Thirring coat

The Thirring–Lense effect has guaranteed my father a permanent place in physics. The problem was introduced by Ernst Mach, a critical spirit. He didn't want to believe anyone, not Boltzmann about his atoms, not Einstein about his theory of relativity and not God about His existence. (There is, however, still some controversy about whether Mach truly rejected the theory of relativity.) Despite all this, his ideas were highly fertile. He posed the question of who actually determines what rotates. At first it's only possible to perceive relative motion and if two things are moving uniformly relative to each other, then either can say it's resting and the other one is moving. With rotation it's different. When you rotate, you can feel the centrifugal force. Mach wanted to know where this comes from; if I were alone in the world, then the statement that I rotate would be meaningless, so why should I feel something? Mach made a false assumption here, because he did not acknowledge the existence of space as a real entity, but still such questions can be useful. Mach came to the conclusion that the fixed position of stars are determinate for us and when we rotate we do so in relation to them. My father's idea was that the fixed stars aren't gods, so what they can do we should be able to do, too. They are simply very heavy, but very far away. Perhaps smaller masses have the same influence because they're closer. As a referee in this battle of wills, my father chose Einstein, whose theory of relativity had to be able to settle the question. He solved Einstein's equations for the following situation: a wooden cylinder is rotating; the fixed stars are at rest far away. The question is, when do I feel centrifugal force inside — when I rotate relative to the cylinder or relative to the fixed stars? Who carries the space time? Einstein's answer was a compromise, but one in which the stars have so much more influence than the real wooden cylinders that these can be packed up and put away. So my father said that there had to be something massive that rotates, so let's start with Jupiter. Jupiter also has plenty of moons that can tell you what's rotating. But then my father got a little nervous about the complexity of the problem and the length of the calculations, and looked for reinforcement. This he found at the Math Institute in a very talented assistant by the name of Joseph Lense. They rolled up their sleeves and calculated the influence of Jupiter's rotation on its moons. The result was devastating. It was so tiny that it would never be visible. This setback demoralized my father so much that he not only abandoned this problem, but basic research altogether. Then something happened that my father had never expected. The Einstein field equations became a cornerstone of physics and the Thirring-Lense (frame dragging) effect one of its most sensitive tests. Then came Sputnik, Nasa and unimaginable amounts of money for research. The experimental verification of the frame dragging effect became the most complex basic research experiment of Nasa's "Gravity Probe B". It took more than 40 years (under the same management), cost untold millions of dollars, and an army of physicists owe their doctorates to their work on it. Today we know that this effect exists. Space and time are dragged by the heavenly bodies just like a philosophizing physicist imagined during the dying days of the Danube monarchy. What it could be good for is just as unpredictable today as it was then unimaginable that there could ever be evidence found that proves it exists.

My Wife

Fig. 7: Peter, Helga and Klaus Thirring from left to right

My wife is a typical mix of the Habsburg empire. When I met Helga, she was a Greek national whose native language was German, originally from Poland. Indeed in reality it's even a little bit more complicated. Her grandfather was a Greek carpet trader, who wasn't from Greece, but from Istanbul. He fell in love with Helga's grandmother on a business trip in Bielitz (Bielsko), which at that time was still part of Austria. A different country and a different religion; but these were not obstacles to get married, this was true also for her parents. He was Orthodox, she was Protestant and was also actually not Austrian. One side of her family came from Wurttemberg, the other were Huguenots from France who had emigrated East. However, with the collapse of the monarchy, tolerance between the different nationalities also disintegrated. Helga was made to experience this first hand. Once she went for a walk with her little brother and was speaking German to him on the street. A Polish man overheard her and came over and slapped her on the face. Naturally, this kind of thing came to an end after Hitler occupied Poland. The situation was not completely cleared up for Helga though, because she was Greek, and Germany was fighting Greece in the war. So again they made her feel that she had the wrong background. Bielitz isn't far away from Auschwitz, and she had to witness how the SS forced rows of Jews into the concentration camp. Then the Red Army rolled into the West, and three days before the Russians invaded, two aunts fled with Helga and her little brother. Her mother had died after her brother was born from childbed fever and her father remained in Bielitz because he wanted to try and save something from his business. However, he was unable to do so, and later he died under unexplained circumstances. After a long flight — it took them days to reach what today would take a few hours by car — they, along with thousands of other refugees, reached Vienna in February 1945. Their sole belongings were what they were able to carry with them in two suitcases. Despite these horrible events, my wife looks back with fondness and almost a sense of nostalgia to the first 15 years of her life.

The reason for this is that whenever possible she spent her time at her uncle's estate in Friedrichowitz. This is near Wadowice, another village that no one would ever have heard of if it weren't the hometown of John Paul II. When I attended a meeting of the Papal Academy of Sciences at the Vatican, we were introduced at an audience with the Pope. She explained to him (in Polish) where she had spent her youth and he simply beamed and didn't want to let go of her hand. The estate must have been a bucolic idyll. When she went with her father they were picked up by a horse drawn carriage at the train station and then they were taken a half mile along a tree-lined street to the homestead. The estate house was on up a hill. There Helga was treated like a princess. Her father did the bookkeeping for her uncle for free, which might not have been such a difficult job, because the properties were largely self-sufficient. They didn't have to sell much because they didn't buy much. The workers were Polish and received board and lodging on the estate. There was very little money. The hierarchy that governed the estate is hard to imagine today. When the farmhands came to Helga's aunt to receive their orders, they had to kiss the hem of her dress. There was quite a scene when Helga sat down and ate sauerkraut soup and beans with the workers in the kitchen. She was told that she had gotten lice there and she had to undergo an uncomfortable delousing procedure to break her of this habit. During the war, Helga's uncle was a strong champion of his Polish workers and was able to save them from deportation. However, this didn't make any difference for him in the long run. When the Russians came, the estate was seized and he was dispossessed. Although we were back in Poland a few times much later, Helga never wanted to see Friedrichowitz again, because that would only destroy her memories.

As she found that the land of plenty was not exactly waiting for her in Vienna, the first thing to do was to make some money as a tutor and to struggle through high school. Without the sacrifices and support of her aunts, Helga and her brother would never have made it. When once I went to see her, I saw that she was living in a very humble room in a back courtyard. She had the Viennese tradition of the Tröpferlbad — a public bath — to thank for the fact that her appearance did not suffer. She could afford to go once a week. Despite everything, she went on to university and majored in English language and literature. We met at the Protestant student union. At first, I wasn't aware of her difficult circumstances. She never complained. Much later, when we bought a house in the country with almost an acre of garden, I could see how much good the wide open space did her.

In our marriage I could save her from material want, but I couldn't offer her a quiet, easy life. She likes to remind me that she had to move

15 times. But she was more than capable of meeting this challenge, and we can look back at our 55 years together with gratification and fulfillment. There were critical moments, such as when I collapsed from cerebral hemorrhage at the Budapest Opera. Helga did not become hysterical over my lifeless body, but remained determined and calm, and although she doesn't speak a word of Hungarian, she found a doctor who made the correct diagnosis. He also knew where there was a hospital with computer tomography where an emergency operation was performed less than an hour later. The next day a medical plane flew me back to Vienna, and after a few months I made a full recovery.

As both of our sons have leadership positions in Austrian industry and economy, the Thirring line will continue in Austria through our four granddaughters. The dreams Mathias Türinger had when he fled to Austria have proven to have been fulfilled.

Fig. 8: Tamara, Sofia, Tanja and Laura from left to right

Stolen Youth (1938–1945)

We had left off my personal fate with Max von Laue's dark prophesy, but everything turned out to be even worse. Our family was to feel the effects a year later. I can still see my parent's troubled expressions as they heard Federal Chancellor Schuschnigg's farewell speech on the radio which ended with "God save Austria" after which our country ceased to exist. Our house was surrounded by military transporters by the next day, and to the question about what does the inscription W. H. (Wehrmacht Heer, the German army in World War II) mean, we received the laconic answer "we're here".

In the "United" Third Reich

To help understand my perspective on the following events, I should first give a rough sketch of my spiritual and intellectual environment at that time. My parents were strictly against the Nazi regime and there was no love lost on the other side, either. My father was removed from his position as a university professor by March 1938, and was sent into retirement. We had to move out of our apartment, which had been provided by the university, and the Gestapo was a regular guest. They immediately confiscated my father's correspondence with Einstein. After the war, individual letters from Einstein to my father appeared and were bought by the Republic of Austria in an auction at the Dorotheum, the leading Austrian auction house. So at least they still exist, but a not insignificant price was paid to the dubious owner.

Another item confiscated was a magazine called "Das Neue Licht" (The New Light) which was about parapsychology. The Nazis were suspicious of this, as were others. When my father was once asked why he bothered with something so unscientific, he replied that it was better to make a fool of oneself once, than to have to be reproached for having shut oneself off to an important phenomenon. However, once when I asked him if anything convincing had ever come out of table turning, he had to say no and maintained an agnostic attitude.

As his pension did not provide enough money to support a family, my father had to eke out a living by taking on consulting work. This meant that we never had a lot of money, but this didn't matter during the war, as there wasn't anything special to buy anyway. Still, my father received relatively good treatment. The father of a schoolmate of my brother, Hans Tuppy, was arrested immediately after Hitler's invasion and was never again to leave the Gestapo jail. The only thing left for his widow to do was to pick up his bloodied clothing. But unlike my brother, Hans survived the war and is one of the most prominent Austrian scientists today.

The high school that I was attending at that time was the Neulandschule, a Catholic school on the outskirts of Vienna. It was considered to be progressive, and in fact as a Protestant I was never discriminated against; I simply went to a different religion class. However, it was very socially aware. Almost half of the pupils came from poor families and would not have had any other chance to receive a higher education. Not only did they not have to pay school fees at the Neulandschule, but they also received a free lunch. This was financially possible because the professors and teachers were idealists. They were not paid for their work, but were reimbursed for their living expenses.

My parents had chosen this school because as day students we could run around in the Vienna Woods for a few hours every afternoon. The school was open to the lower classes, so it was not considered to be an elite school. To the question as to why I didn't attend an elite school, since that's where one would meet those people who would be "useful" later in life, my father arrogantly replied that we did not have to stoop to such things. In fact, there were pupils at the Neuland from all walks of life, and even the nobility was represented. We even had a Habsburg, we liked to ask him if he'd just come from the cleaners (Habsburg was not only the name of the Austrian Emperors, but also the name of a large laundry). There was a certain team spirit in our school and all of us, pupils and teachers alike, were on familiar terms.

Despite its address (Alfred Wegenergasse 8–10, named in honor of the founder of the theory of continental drift) the school did not always welcome the latest advances in science. Darwin's theory of evolution was rejected on the grounds that "a worm can't become a bird." There were also some absurd scientific claims made by some Catholics at the school. I once saw a pamphlet propagating the hollow earth theory. The explanation was that a hollow earth would make it easier for Jesus on Judgment Day, as He could then appear at the center of this hollow earth and simultaneously be seen by the whole world. Although geometrically correct, this idea belongs to another world than where the scientist, who values truth above all else, questions every assertion ten times before putting it to paper.

Unlike in my elementary school, where about 30% of all of the schoolchildren were Jewish, there were no Jewish pupils at the Neulandschule. In my school days, anti-Semitism among the kids was unknown. Jews were just those kids who went to a different religion class. My best friend in elementary school was Jewish, and it never mattered at the parties at his house if you were Jewish or not. Later in the Neulandschule I also can't remember hearing an anti-Semitic remark from either the Catholics or the Protestants. This doesn't mean that there weren't ethnic stereotypes; for some, the archenemies of all Austrians were the Italians (referred to pejoratively as "Katzelmacher" or pot maker, presumably referring to some Italian immigrants working as dishwashers).

This went as far as the food: "We'd rather eat our hard sour Styrian apples than that Italian junk."

The Neulandschule also did not place a general liability on all Jews for the death of Jesus Christ. When directly posed this question, Rev. Dolezal, who would later be very important for me, replied that "Naturally, the Romans were also partly responsible, but the Jews bear the brunt of responsibility, just like Luther was partly responsible for the church schism in Germany, but the Catholic Church bears the brunt of responsibility." Put in more general terms, this means that a rigid religious establishment is what is mainly responsible for martyrdom.

How did the German occupation change the school? At first, not a lot, except that some teachers started walking around with Nazi party pins the next day. Others who disagreed with the new government simply disappeared, and they took the school's spirit with them.

Our classroom was Germanized in the sense that now pictures of how the ancient Germanic people lived hung on our walls. This wasn't so bad; I actually really liked some of the blonde Germanic girls. However, I was soon bothered by a certain contradiction in the logic of what we were being taught. At first we were told that Germany was in no way involved in the Spanish Civil War, but when it was over, the Condor Legion were celebrated as the victors when they returned. In Munich, Hitler declared (and signed on paper) that he would make absolutely no more territorial claims, and then Czechoslovakia was annexed piece by piece.

Still, no questions were asked about these events, apparently many were already by then suffering from memory loss. We, however, were spending six periods a week going over this hypocrisy. We were reading Caesar's account of the Gallic Wars in Latin class, where he presents each conquest as inevitable. The similarity to Hitler's propaganda about the war which was forced upon him was unmistakable.

The further changes in the Neulandschule took place gradually. Our beloved afterschool teachers were replaced with no one's favorites, some teachers were never seen again and others simply appeared; one in particular, a new school director.

I personally felt the effects of the new direction in the following way. In the Neulandschule, some of the kids had gotten together to form a group that took trips together and met up on other occasions. During these meetings Slavic folk songs were also sung, which I still prefer today. After the Anschluss, naturally such groups became illegal, but we still stuck together.

Once as Sergej Jaroff and his Don Cossacks was appearing in Vienna, it was clear that we would all meet at the concert. We practically filled one standing room area, and our applause ensured countless encores. We were all startled to find that the area had been blocked off as we started to leave and go home. Luckily I was quite small and could slip through a crack. My bigger friends were arrested and taken to the Gestapo headquarters to be interrogated. They were shocked to see our new school director coming and going around midnight. He apparently held a leading position there.

Although my friends were soon released from jail, we couldn't continue our lives as before. Finally, even the school building itself was requisitioned and our Neulandschule ceased to exist. We were forced to go underground, as I will explain later.

With the new teachers and rooms, the general standard continued to go downhill. Although some scientific subjects continued to be taught well, as they were not infected with ideology, they way other subjects were taught would seem grotesque today.

For instance, we learned in German class that only the Führer speaks the best German. Hitler's style might have functioned on a demagogic level, but was linguistically repulsive, common and primitive.

In music, Johann Strauss became a persona non grata for having written the Gypsy Baron which glorifies a particularly inferior race, the gypsies. In geography class, we learned the exact value of different peoples. It was said that a German was worth as much as 20 Polish people or 80 Russians. We never found out how large the negative value of Jews would be; apparently this was too much for the geography teacher to calculate. During this period of intellectual squalor, the Rev. Dolezal I mentioned before came to our rescue by reviving the old Neuland community in a new form. He was a brave man, because as I will shortly describe, a secret meeting could easily end at the gallows. He demonstrated his bravery again at the end of the war when he was able to hide the female members of our group (which was now mixed) in his church, so that they were safe from being raped by Russian soldiers. He had hid people there the entire time, which was twice as dangerous. What could be done during the air raids? You couldn't take them with you to the air raid shelters, where there were always informers. You also couldn't leave them above ground, as that was exposing both them and himself to great danger. If the church was hit, the hiding place would be found. It didn't matter who those people were, Jews, opponents of the regime, whoever, in any case they were illegal and we kept our silence. He hid a friend of mine, Alfons Stummer, after he deserted in February 1945 as a conscientious objector. Alfons saw what could become of him, as from his hiding place he could see how the SS troops dug through basements and murdered the illegals they found on the street with a shot in the neck.

We met once a week in his rectory and discussed religion and politics. This is where I got to know great works of the past such as the St. Matthew Passion by J. S. Bach, and where I found out about the horrible present, with its concentration camps and the Holocaust. In the meantime the war had begun, and as the Germans' final victory continued to be pushed further and further into the future, more and more of us were drafted, and these meetings also were extinguished.

In the Army

One day at the beginning 1943, it was just after Hitler had lost the battle of Stalingrad, all of the boys from my class were summoned with their parents to the screening room of the "Auge Gottes" cinema for a speech from a high-level party official. The long and short of it was that we all now had to join the army, as a part of an anti-aircraft artillery battalion. Not even quite 16, my youth was now over. The party official found some comforting remarks as well. In Stalingrad it's true a regiment from Ostmark (Ostmark referred to Austria as Austria had ceased to exist) was completely decimated, but he had the confidential information from the Führer's headquarters that it should be reformed under the name of Hoch- und Deutschmeister (the name of an historic Austrian elite troop). Then the guys from the Altreich (the Old Reich, or Germany) could no longer look down on us as "Ostmark wimps".

You can easily imagine just how comforting this thought was for our parents and for us, whose sons and friends had starved, frozen to death, and come to a miserable end in Stalingrad.

In this instance, I was granted a short reprieve thanks to a fact that I sometimes thought was more of a bother, namely that the letter "T" appears almost at the end of the alphabet.

First we were taken to the base to undergo basic training, and learn things such as how to march. To strengthen our constitution, we were vaccinated against everything possible, including scarlet fever. I didn't think I needed it, because I'd had scarlet fever and they said that you couldn't get it twice.

But then the unexpected happened. A few days after the immunization, red rings developed around the area where I had been vaccinated, I developed a high fever and was diagnosed with scarlet fever. The doctors explained to me that the serum wasn't very good and still contained active germs. Apparently the bottle hadn't been shaken well and the last few got an overdose from the dregs, so that I and a few others from the end of the alphabet were infected. Scarlet fever was a serious illness at that time. There were no antibiotics. It meant six weeks in hospital with a two week individual quarantine.

This was quite boring, but at least I could read and brush up a little on my withered education. I also knew that scarlet fever often caused heart defects. I cultivated this to my best ability so that I was confined to the hospital and declared an invalid until the fall of 1943. I was able to attend (a completely different) school for a short period and was also able to hike through the mountains in Western Austria with the Dolezal group in the summer. The group had now shrunk to include almost only the female members by this time. I didn't care about this, but it meant that the few roosters in the henhouse had to carry most of the supplies.

We were able to spend some glorious days, but I knew that I couldn't get away with faking heart problems a second time. So it came to pass that starting in the fall of 1943, I spent a year on the Nussberg mountain near Vienna, where my unit had been transferred.

From the top of the Nussberg there was a fantastic view of Vienna, and on mornings when the sun was shining, the Danube split the city like a shiny ribbon. There wasn't any air pollution back then, so that on clear days you could see the small Carpathian mountains in the distance. Therefore we couldn't complain about our location, and were also in other things more or less well accommodated. The toilet, actually a latrine serving 40 people, was a wooden shed built over a cesspool with a board in the middle which had some round holes cut into it. There was also a beautiful view from here, but it was a bit drafty during snowstorms. We didn't really care about this, though, and after we had become masters of the bedbugs in the barracks, we actually found it quite comfortable.

There was even rudimentary schooling for us, and because nothing much was happening yet in our air space, we spent the rest of the time cleaning grenades, shells and canons. From conversations with my fellow recruits I knew that about a quarter of us children serving as Luftwaffe support personnel were strictly against Hitler and about a quarter strictly supported him. The political orientation was presumably similarly split amongst the regular soldiers. One sergeant kept admonishing us, "Guys, we've got to win the war." He was a teacher from East Prussia and apparently understood that for him, a lost war meant a lost home. He could also feel that I did not share his convictions and certainly he was suspicious of me, but I didn't suffer any consequences.

All told, our life in the army was almost idyllic compared to the terror that reigned in some civilian areas. I'm still appalled when I think about the new true believer school principal. Two pupils had listened to a foreign radio station and were rash enough to write down the news and distribute it in class. This was discovered; they were sent to a military court and sentenced to death. Their parents made an appeal for clemency, which was granted due to their youth.

Now the principal felt his Germanic honor demand that he appeal their pardon. This appeal was approved, the pardon rescinded and the kids were beheaded.

With the outbreak of winter, life on the mountain became monotonous, but then something happened which brought me a lot of credit I didn't deserve.

Once when I was on phone duty, it was announced that the German Luftwaffe ski championships would be taking place in Zell am See, and qualified participants should be registered. Skiing has always been (and still is) my favorite sport, so I was very tempted, although I didn't think I had a chance to win anything.

Luckily the head of our company was well-disposed toward me, once I was even allowed to play something on the piano for his daughter, and he nominated me. The famous ski jumper Bradl — he was the first to jump over 100 meters — was in charge of the championships and managed them with a great deal of care, but without any technical equipment. There was a snow storm and it was impossible to prepare the slopes or to postpone the contest. I tried to do my best, but with the poor visibility and the unrecognizable slope, I ran into a fence. As I fell, the material with my starting number twisted my 18, so that it looked like 16. Indeed, the guy who actually started two numbers ahead of me didn't fare much better; he fell and his 16 was read as 18. But because he reached the finish line long before I did, the number 18 received a fantastic time, and I won the youth title. There were some other irregularities, though, so that when everything was re-examined and I fell back to third place.

However, by the time this was clear, the camera team for the Wochenschau, the weekly news reel, had already left. I was thus presented as the winner of the youth title before every movie shown in the Third Reich.

Even my brother, who at that time had wound up in Ukraine with the army, was surprised to suddenly see and hear his last name in the news.

Our situation began to worsen in the spring of 1944 simply because of problems with the basic staple of the Prussian diet, the potato. We had received a large shipment at harvest time, and had to store them in a so-called potato clamp over the winter. This was a large ditch into which the potatoes were poured and then covered with straw and dirt. We had apparently underestimated the winter as they were frozen and began to spoil come spring. To get at the potatoes that were still good, we had to dig around in the mud with our bare hands like pigs, looking for anything intact. But beyond these outward circumstances, I had the feeling of wasting the best years of my life without really learning how to do anything properly.

Therefore, when the announcement was made that they were looking for people to receive technical training for what is now called the radar unit, it seemed to me that that would be a better choice and I volunteered. I was detached to the Westerwald forest and I was looking forward to the trip. I would be going through Frankfurt and could finally see the city of Goethe.

I was completely shocked to find nothing but smoldering ruins. I had arrived on the day after the devastating air raid which had completely leveled Frankfurt. My guardian angel had chosen a good itinerary, because if I had arrived a day earlier, I would probably not be able to be writing these lines. Changing trains for the Westerwald was made much easier, though. I needed to go to a distant train station, but I could easily see it directly as there were no buildings left standing in between.

I'd only heard about the Westerwald forest from the well-known song and it seemed like paradise compared to Frankfurt. The area was lovely, there was no cold wind blowing, the food was much better than in Vienna and our posts were very spacious. Everything seemed fine, but then I got wind of what was really going on. It was announced that after our training we were to become part of the Waffen SS. I knew immediately that I had to get out of there, but how?

Luckily we still had to take an entrance exam to prove our math skills. This is where I saw my chance. I had worked on high school math material enough to have a good overview of it. I was even in the position to tell the teacher when he asked, "This right, Thirring?", "Yes, sir, it's actually right!" after a long equation.

I could now use this to my advantage. I was able to see where in the exam problems I could make mistakes so serious that I would fail, but not in a way that would be obvious sabotage. The SS didn't like sabotage.

I'm still proud of the fact that I was able to successfully walk this tightrope and was sent back to Vienna. My old company didn't care that I was a failure in math; after all, I had already proven my athletic abilities.

The summer of 1944 was heating up, particularly in the air space surrounding Vienna. The Americans had become so well established in Italy that their bomber squadrons could penetrate through to Vienna and air raids were a fact of daily life. We were deployed and for the first time in my life I was faced with the dilemma of having to shoot at people. I had always taken the standpoint that this was Hitler's war and I wanted nothing to do with it, least of all shoot at people. But now how could I avoid it?

We had the following duties. One of us located the bomber in his binoculars, I read the degree coordinates and calculated the correct allowance and passed this data on to the gunners over a microphone, where the data was used to make the constant adjustments necessary and then they fired about every ten seconds. My tactic was to recalculate the data so that the shells would explode about 50 meters behind the airplane. They couldn't do much damage any more, but it still wasn't so obvious that something was wrong.

This hardly decided the outcome of the air battle over Vienna — with the thousands of shells spent, there were around 2–3 bombers shot down during each attack. These were more like random hits and I didn't have the feeling of being involved in murder.

In the course of these air battles I noticed certain discrepancies in the foreign reporting on them. We had leave once a week and I could visit my parents. Despite the risks involved, my father was an avid listener of foreign news reports on the radio, and I diligently listened with him. Because everything they said on our radio was a lie, I thought that everything the others said had to be true. Here, though, they were saying that military facilities were being bombed in Vienna, but I could see from our positions exactly what they were bombing. This included fuel depots which were easily recognized by the black clouds of smoke, but also purely residential areas.

At the end of the war when it could be seen what was left, everyone could see for themselves that not only military or war-related facilities had been bombed, because the Opera house or the Burgtheater would hardly be called military.

We were the actual military facility, the anti-aircraft posts.

But we were never bombed; the only danger we were threatened by came from ourselves. The shells we shot up did not stay up, of course, and shrapnel came raining back down on us, whistling. We were never attacked though. I assumed that we weren't worth it for the bombers, but were more of a target for the strafers, the low-flying jets.

These were rare, I only spied one once. I followed it with my binoculars to see if it would attack an anti-aircraft position, as I knew of course exactly where these were. But this was not his intent, he stayed away from these and was having fun hunting down people and popping them off like flies. I was so outraged by so much cowardice that I thought shooting down people like him would not be doing humanity a disservice. From then on, I stopped taking the trouble to recalculate the angle allowance.

The rest of the summer of 1944 brought all sorts of uncertainty. On July 20th first there was the exuberant rejoicing after hearing the news about the assassination of Hitler, then the deep despair when it turned out to have failed.

Then the authorities decided that we were now big enough to be sent on a work brigade to Poland. Our last report card (from the 7th grade of middle school) was to count as our high school diploma.

That was where the Red Army was advancing, and work duty in Poland was not a pleasant prospect. My father told a former medical colleague, Dr. Oppolzer, a surgeon, about his worries, and he said that he would be able to fix things. He had just performed an appendectomy and had kept the appendix for further study. He could remove my appendix and switch them around so that I would be entitled to three months of convalescence.

This is what happened and not only did it save me from the work brigade, but it also directed my life towards its future course in that I received the fateful letter from my brother. He had always been considered to be the big genius while I was the little hanger-on. There was no question that he would be the one to follow in my father's footsteps while I tended to a career in music. As he was older he was called to the army earlier; he was wounded often, but had always managed to stay alive.

Now I read in his letter from the Eastern front that he was certain that he would not survive the war and that it was now up to me to continue the scientific tradition of our family.

Sadly, he was proven right and I took his instructions seriously. I went to my father and asked him for a textbook on theoretical physics. He gave me the textbook written by Joos, and this was the right choice.

It had something over 600 pages and I had figured out that I would be able to finish it during my three month reprieve when I managed seven pages each day. I threw myself into it and was able to complete my workload. As we will see, this was crucial for my future career.

However, nothing lasts forever and also the appendix story couldn't save me until the war ended. I was called to the mountain infantry in Landeck, but I didn't stay long.

We had a very experienced instructor who had survived numerous battles. One day (February 19th) he wanted to show us how to use a rifle grenade even without a rifle. Rifle grenades are those small grenades that fit perfectly in your hand. When you pull the plug back in there's a mechanism that postpones the explosion for just a short time. This means you've got to throw it as far as you can as fast as you can and then duck. This was the theory. In reality what happened was that there was a violent explosion as soon as our instructor pulled the plug. He was torn to shreds, I was thrown back 12 feet and altogether there were 12 dead and 30 wounded.

The explanation came later that the production had been sabotaged and the delay mechanism was set incorrectly. The use of sabotage, which I had also supported, almost cost me my life, but in actual fact ended up saving it. The rest of the company which had not been badly wounded were sent to fight the partisans in Yugoslavia and were not heard of again, while I spent the rest of the war in an army hospital.

The casualties were taken to the nearby hospital in Zams and according to the severity of their injuries, had to wait in front of the operating room. When I arrived with a fellow soldier, the medic told me that my comrade had to go first and that I was not a serious case. I was surprised to hear this, because I had blood coming out of where 24 pieces of shrapnel had bored into me and I was covered with blood. The other fellow had just two small cuts in his stomach. The medic was right, though, as a cut in your stomach at that time was a death sentence. Antibiotics hadn't been invented yet, and a dirty germ in the abdominal cavity would automatically lead to peritonitis and an agonizing death. He lay next to me in the room, and the inflammation and the pain began the next day. There wasn't any pain medication, so he was screaming all day and all night. Only on the third day did he get weaker and you could hear his laborious breathing in the night, until death finally had mercy on him.

For some reason that I've never been able to reconstruct, just before the explosion I turned to the side. The shrapnel therefore entered my right arm and my right leg, one piece penetrated under my left eye into my skull and came out through my temple, but my abdominal wall remained intact.

I would not like to write much about my time in the military hospital. In the long weeks you hear and learn all sorts of things about your fellow human beings, starting with perverse bordello adventures to reports from concentration camp guards about the conditions in the camps. I will return to the conclusions I made from these atrocities in the epilogue.

To my great pleasure, as soon as I was able to be moved, my parents were able to arrange to have me transferred to the military hospital in Kitzbühel, where my parents had their second home. I was able to wait out the end of the war in peace, but at the very end I almost came to share my brother's fate. The last message from him came from Frankfurt an der Oder and remained his last. In my reply I wrote about how I'd met some idiot in the hospital who still believed in Hitler's final victory. This, of course, was extremely risky. He never received this letter as Frankfurt had long been taken by the Russians and letters addressed to missing people were repeatedly opened and read. Exactly towards the end, comments suggesting insubordination were punished with particular severity. The commander in Tyrol, General Kesselring, was still ordering many people be shot even in the last days for similar things.

When the commander of the military hospital came to me and said that he had received a notification and I was to report the next day, I remembered the letter and felt the fear rush through my body. I have never found out what was in the notification. The next day the Americans marched into Kitzbühel and I escaped, never setting foot in the military hospital again.

The Horrors of War

I can only formulate my denouncement of the crimes of war with much deliberation. Too many memories are running through my mind. However, I don't want to rely solely on my memory, which can already be a bit faulty. I think a passage from one of my brother's letters will suffice. The yellowed paper is lying in front of me and I am copying it word for word.

"Together with a group from the Ustascha legion[b] *we were supposed to surround a group of partisans in a village and take them prisoner. We had*

[b] Croatian Nazi collaborators.

about 100 prisoners, from almost every walk of life, and were searching them and interrogating them when a big, broad shouldered Ustascha sergeant came up to us and asked for about 10 prisoners as he had a special assignment for them. The man had a bloated face like a bulldog and his eyes were hard like steel; they were cruel. Still we fulfilled his wishes and he chose 10 younger guys, between 15 and 17 years old, and tied their hands behind their backs. Then he took them to the side, suddenly pulled out a dagger and slit the first guy's throat, fumbled around a bit until he'd found the artery, bit it off and sucked out the blood. His eyes shone like he was intoxicated with madness and his massive body was quivering with lust. His victim grew weaker and weaker and suddenly collapsed; he let him lie there and bleed to death and turned to the next one. We hadn't been paying attention to him at first and by the time we noticed what was going on he already had his 5th victim in front of him. We were so paralyzed by this horrific scene — four guys were lying with slit throats softly moaning and trembling on the ground and he leaned over the fifth like a giant demon and sucked out the blood with a diabolical greed — that some had to run away, they couldn't take it. We had already been up against some tough characters and had banned pity from our souls, but still in this situation we all agreed that this monster must be shot. However, some other Ustascha members appeared and prevented us from doing so. We had to watch how he sucked the blood out of them, one after another, while they were still alive. He had so much practice that not a drop of blood was spilled; he must have done it often. The victims tried to get away, but their resistance was useless against the sergeant's overwhelming strength. It seemed like he only liked it when they would still struggle a little more. We finally put an end to the horror with a couple of hand grenades. Perhaps it's no coincidence that the sergeant was severely wounded by two pieces of shrapnel in his throat, while his comrades ran away. We let him lie there amongst all of those he had slaughtered until he bled to death two hours later."

The APO letters from my brother fill many binders and they appear to shout out to the world and tell about what has slipped away to being forgotten. I can only summarize what the war taught me: a person's moral framework must be constructed carefully. If the right role models aren't found at home or elsewhere and if the Ten Commandments are never learned, then the framework will remain just a crumbling crust. In the hands of demagogues it will turn into a thin skin which can be deformed in any way, shape or form and when it rips, a beast can emerge. In war, when a human life is worthless, these satanic powers have an easy time of it.

How can anyone justify such outrages?

Training and Education (1945–1959)

My memories of May 8, 1945 still shine brightly. I had gone up a mountain and was lying in a meadow, listening to the bells ringing in the valley. They were ringing in the peace.

The horror trip of the last six years was over. We could once again do what we pleased. At the same time, it was clear to me that we were standing before ruins, and it would require endless amounts of stamina and patience to rebuild the country and make it inhabitable.

The first thing I had to do was to legalize my status. Whoever had served in the German army was per definition a prisoner of war, and if he had not been captured, he had to report voluntarily to be released from a POW camp. I chose to go to the Americans, for although they were somewhat rude, they had soup three times a day. Then we were transferred to the French. They weren't rude, but they also didn't have any soup. I was released just in time. Although I had become skin and bones, it was still before the French realized that the POWs could also be used as cheap labor. The rest of my fellow soldiers had to work in French coal mines for a couple of years before they received their official release documents.

After I was able to legally think about life as a civilian, I had to think of how I could make it work. I could not completely rely on my father, because he couldn't continue his consulting work in Kitzbühel. He had been re-hired by the University in Vienna, but it would take months until anyone saw any money, and there was not much of it at that. The state was still taking baby steps towards reconstruction and there was almost nothing of value for it to build on. My father came up with a good idea. He said that because so many men were POWs, there would be a labor shortage, so I should go work as a farm hand. I could then start going to university in the fall. That is how I got to know this kind of work as well and it helped me get by when the times were so bad. My duties were to get up at 4:00 a.m. and scythe the field until 10:00 a.m. This was naturally all done by hand; not only were there no machines, some of the fields were so steep that they couldn't have been used anyway. There were Tyrolean dumplings and salad to eat every day, good healthy food. We ate this outside in the fields. Just once did the farmer invite us to eat inside, then we were given pancakes and cherry compote. There were no plates, and we all ate from the same bowl, which caused me a bit of a problem when it came to the compote. The cherries had pits, and I didn't know what to do with them. It probably wouldn't be healthy to swallow them all and I didn't want to just spit them back out into the bowl like everyone else was doing. So I laid them on the table, as there wasn't a tablecloth to get dirty and the table had wormholes anyway. Then the farmer said to me, "What kind of a pig are you, laying the pits right on the table?" I learned that cleanliness was also in the eye of the beholder.

There was no money to be earned, but at the end I got something much more valuable; a sack of potatoes.

Then it was high time for me to return to my education. First I had to close some of my gaps as my high school education had been run over by a tank, so to speak. On the other hand, I didn't want to lose much time, as I had already wasted almost three years. The question was how. I was stranded in Tyrol and didn't have a diploma. I couldn't even show a report card from school because these were all in Vienna, which at that time was unreachable from Tyrol. What helped me was the coincidence that a physicist, Arthur March, was the dean of the University of Innsbruck and I was able to convince him that I had at least mastered the material of high school physics. He overlooked the fact that I didn't have any college prep exams and let me enroll. I'm still grateful to him, because he saved me almost two years, two years which would have prevented me from getting to know Albert Einstein.

So I began to go to university, first in Innsbruck for a year. Luckily, one of my father's cousins lived there and I could live with her, at first in the basement and then a room became free upstairs. The main problem was finding something to eat, as the ration cards were for minimal amounts. I was fortunate to have my sack of potatoes and for the fact that I didn't mind eating cod liver oil. For whatever reason, this was available, but no one wanted it. Potatoes roasted in cod liver oil were very healthy. This was monotonous over time, but fortunately I was able to find a big piece of bacon which made for a delicious change.

This didn't last for long, however, because I had to leave my basement window open at night to get a little air. No one could get in, but one night I saw some shadows darting in and I could recognize it was a cat with my bacon in its mouth.

All in all, I remember the time in Innsbruck fondly. The next year I was allowed to return to Vienna, and with my report card from Innsbruck, I could study at the university there. I was a very hard worker and finished all of the requirements in three years, the minimum amount of time. Then I was ready for my PhD. This was actually a bit faster than was permitted and I have to thank my advisor Felix Ehrenhaft for this. He was an unconventional man and somehow I managed to become one of his protégés. (There will be more about him later.) We hadn't set up a time for my defense, or even agreed on a topic, and he saw me sitting in the hallway and said, "Exams are like going to the dentist. It hurts the least when you don't expect it. Come on in and let's get it over with now". I knew that I had to imitate his style and every reply was a bold statement. I quoted theories I knew he would not be familiar with as an experimental physicist. This impressed him and I passed with honors.

But then the whole sham was exposed. Officially, I needed to go to an office and present a copy of my diploma. I stood there empty-handed. Once again Providence was smiling down on me and someone in the room said that there was a special decree for people who didn't receive a diploma because of the war. If they still passed all of their requirements and their defense, they were eligible for a PhD. I clung to this sheet anchor and it held; the formalities of my academic career were successfully completed.

When it comes to the content of a scientific course of study, this is, of course, never exhausted, and I also had to plow through all of the beginner's mistakes. They're like the childhood diseases that you have to have to build up your immunity. First I came down with what I'd like to call the math measles. This is something everyone must go through, but can be overcome with a little discipline. I was infected because it was easy for me to understand what a formula said, and I was only interested in the final results. The steps in between I considered to be unnecessary weeds that could be held in contempt and ignored. This worked out fine up until the day that I had to derive a result myself, and I stood there at a complete loss. It began to dawn on me that I would have to solve every intervening calculation with precision — ideally, in different ways to find out which is best. Of course this takes more time, but then you learn the different versions of a formula and can store them in your memory to call upon during lectures or discussions, which is essential. The fact that I taught myself how to do this allowed me later to be able to work with the leading people in the USA, because this takes place without any notes right at the chalkboard.

This ability was not enough, however, to protect me from what can be described as physics flu. This is quite a common complaint, but the healing process must be thorough or else it could become a chronic condition. Even when the mathematical correctness of a formula is comprehended, it doesn't mean that the conclusion relevant for physics has been understood. A formula in physics says that certain concrete amounts are equal to each other and you have to be able to conceive this amount to see if the equation says something banal or unexpected. Already when writing out the formula you've got to be able to extrapolate its meaning. For example, if the radius r of an atom plays a role in the result, and it is expressed by the elementary charge e, the electron mass m and Plank's constant h, $r = h^2/me^2$, you need to gather these factors and not leave them scattered around in the formula: it's better to express $\ldots h \ldots m \ldots 1/m^2 \ldots 1/e \ldots 1/e \ldots h$ as $h^2/me^2 \ldots$ If you don't, the formula remains a sterile bunch of numbers without sparking anyone's imagination. But imagination is exactly what you need to develop something like intuition. I will illustrate this using the example of the theory of relativity.

After overcoming these initial difficulties, for me 20th century physics became a glistening playground covered with questions of different complexities. I wanted to contemplate them undisturbed if possible, and when one playground became too crowded, I went to another. There were other cliques there, of course, and they were suspicious of newcomers, but I didn't mind. I often found that they were blocked by problems that had already been overcome elsewhere and then I was in a fortunate position to show them.

The Theory of Relativity

While I was studying up until receiving my PhD in 1949, I completed just one scientific paper and it was the result of my dealings with the theory of relativity. The name alone gives it the wrong aftertaste. A theory is often understood as a shaky intellectual construct and the word "relativity" almost implies meaninglessness. It just refers to the distinction between the relative and the absolute and because the latter is more important, it should really be called the "doctrine of absolutes". During the Nazi period it was degraded as a "Jewish swindle". The special theory of relativity doesn't say anything concrete, just how space and time appear to different moving observers. These are just covers we slip over the phenomena to be able to register them. So it's more of a human convention than something anchored in nature. It gets even more surprising when Einstein fuses it to matter in the general theory of relativity so that they influence each other. Very few people were interested in what these theories said in more detail while I was studying — they were generally viewed as a nice accessory to the photo of Einstein sticking out his tongue.

For me as a student of physics, it was a point of honor to find out what Einstein actually had proven, or had he just guessed something; what exactly did his formulas express and when were they important? In the special theory of relativity they deal with how space and time are connected during the transition to a moving frame of reference, together with the most famous formula in the world:

$$E = mc^2.$$

I will briefly show how I saw this theory to give a better picture of the challenges I was facing at the time. Even when it is rumored to be much too difficult and incomprehensible mathematically, I will try, as not a single word of this rumor is true. I just need the letters x and t from higher mathematics. Still I will mark it as an excursion, and whoever wants to can skip over this educational adventure and won't lose anything in the rest of the story.

In the special theory of relativity there are three geometric effects, the relativity of simultaneity, time dilation and length contraction. These are easy to understand when visualized and illustrated (see below). The comparison between clocks at different locations requires a means of communication and when even light just has a finite speed and you have to make allowances for the time of transmission, there can still be irregularities. When there is already disagreement about what is simultaneous, then the size of a moving body also comes into question, because this is defined by the distance between the beginning and end at the same time. My father illustrated how the size of a body can appear relative quite nicely in his lectures using the theory of calf relativity.

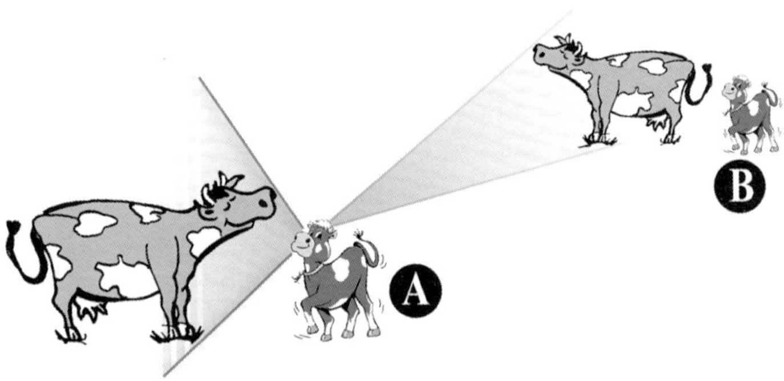

Fig. I: The Theory of Calf Relativity

A young calf naturally identifies the size of an object by the angle under which the object appears. Therefore calf A says to calf B, "My Mama is a lot bigger than yours". But calf B sees it differently as it is closer to its mother and says, "You're wrong, just look, my Mama is bigger". Even though this seems to be a contradiction, both observations are correct, as each calf is closer to its mother and she then appears bigger. This shows how apparent contradictions can be corrected by distinguishing between the relative and the absolute.

Excursus on the special theory of relativity

We will now take a closer look at how the theory's geometric effects are generated, as they have long been a source of irritation for their contradictions in logic. To document the events, we must say what was happening at location x and at time t. Our coordinates are the pair (x,t), and just like for setting up a rendezvous, it takes two determinates, where: x, and when: t. In keeping with the latest trend, we'll call (x,t) an event. The events (x,0) are those locations where the time is 0 (the x-axis) and (0,t) are the events taking place over time at the location zero (its world line, the time axis). The question now is who determines when time 0 is, who sets the clocks at the different places x? The genius of Einstein led him to the best referee there is, light. He said that nothing is faster than the speed of light. It also has a finite speed

c = 300,000 km/s, but we can make the following allowances:

when I want to synchronize two clocks at the locations 0 and x, then I send a ray of light to both sides simultaneously from the middle at x/2. The events when they reach 0 and x have to be at the same time, they're (0,x/2c) and (x,x/2c.) If I want to illustrate this in a drawing, first I have the problem that x represents a point in three-dimensional space, but the paper doesn't have that many dimensions. So we draw just one dimension from space, from event (x,t) x to the right and t going up, keeping in mind that there should actually be three xs.

Secondly, we need to choose a scale to be able to compare space and time. It would be the most convenient to take the speed of light as the reference speed, using 300,000 km as the unit of length and seconds as the unit of time. In these units c = 1 and the world line of a ray of light runs at 45°. The world line of every other body has a speed of v < 1 and must therefore run at a steeper angle.

Now the question is how an observer moving with the constant speed v would map the world. For him, the time axis is the line x = vt and he believes that he is resting on it. To determine the x axis, he uses the Einstein convention, just like the inert observer would because he thinks he is the one who is actually at rest. If he wants to know when the zero hour will be for a fellow traveler, he lets two beams of light to blink simultaneously from the middle x/2, one forwards and one backwards.

To sort our thoughts more precisely, we'll imagine that the first observer is sitting in the back of a train, and the observer at x is at the front of the same train. When we draw this synchronization procedure we can see right away that the events at the zero hour are at t > 0 in the moving train.

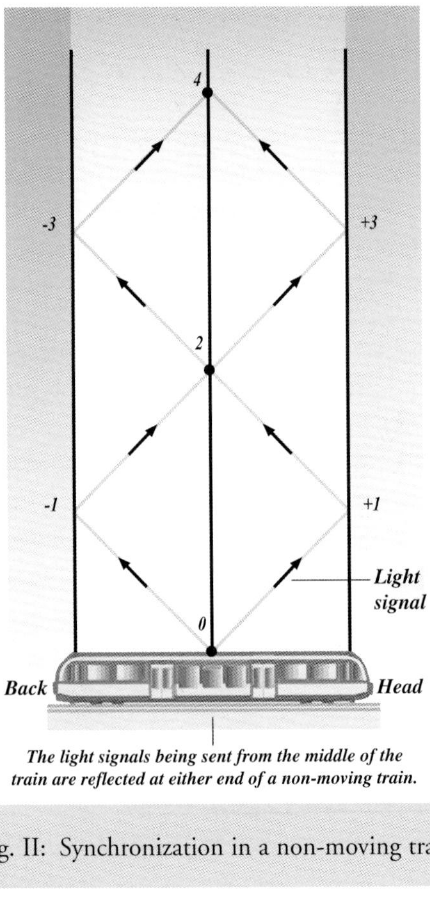

The light signals being sent from the middle of the train are reflected at either end of a non-moving train.

Fig. II: Synchronization in a non-moving train

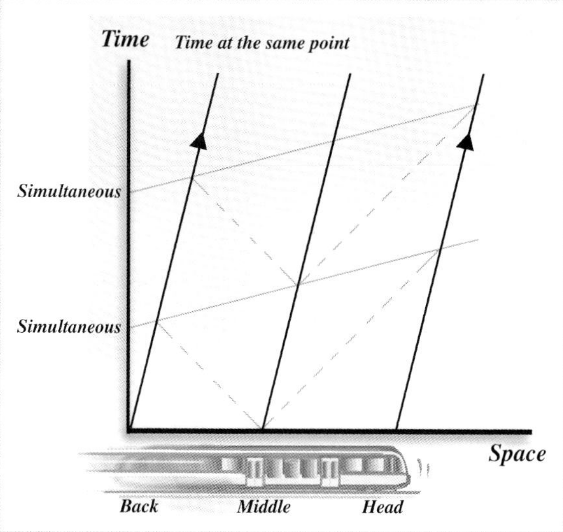

Fig. III: Synchronization in a moving train

So the events at the zero hour are different for an inert and a moving observer. The "inert" observer knows this, because the moving observer behind him seems to be moving towards the light and the moving observer thinks that the inert observer is in fact moving away from it. But who is actually moving now? This is where Einstein made the crucial postulation that there is nothing in the laws of nature which distinguishes absolutely uniform motion. Therefore, each observer can claim that they have the "real" time because they both think they're at rest. The conclusion is that there actually is no "real" time. Sometimes you can read "there is no time", but I would rather say it the other way around, there are too many conventions of time, and none is better than the others.

The discrepancy about the zero hour also causes a discrepancy about which distance is longer. As we are only interested in this contradiction and not in the numbers, let's take a look at another drawing. The parallel lines once again represent the front and back end of a train, it is at rest if they are vertical and is moving in the others. If one distance encloses the other, it will naturally say it's the longer one, but this must take place at time 0. However, it can happen that at one time 0 line one segment is enclosed in the other, and on the other 0 line it's exactly the opposite.

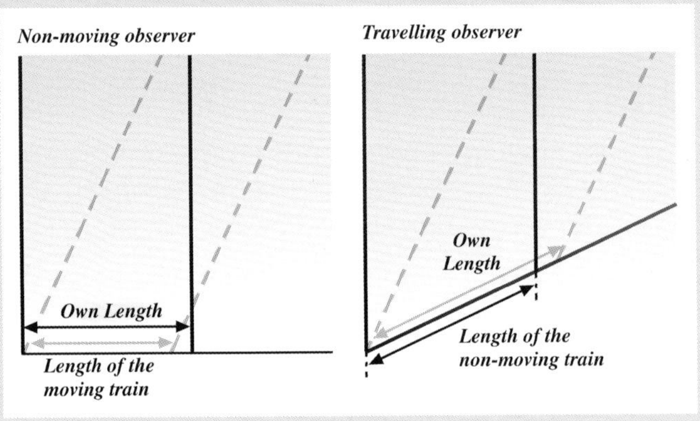

Fig. IV: Length contraction

So you once again have the situation of the theory of calf relativity. If you switch x and t around and talk about the length of a period of time, you can, of course, also generate conflicting views about the time elapsed.

The question now arises regarding how come both observers have the same value for the speed of light despite the distortions in space and time. Let's take a look at why the non-moving observer gets the value of 1 using these units for the speed of light. He sends a ray of light from event (00), from the point of origin 0 and has someone who registers him at event E = (11). They conclude that speed = distance/time = 1/1 = 1.

When the moving observer uses the same experiment, he gets other values for distance and time. For him, the t-axis bends towards the right and the x-axis is moving up. You can see in the figures above that the angle of the t-axis with the vertical is the same as the x-axis with the horizontal. For the moving observer the distance is the way from 0-B and the time B-E (see Fig. V)

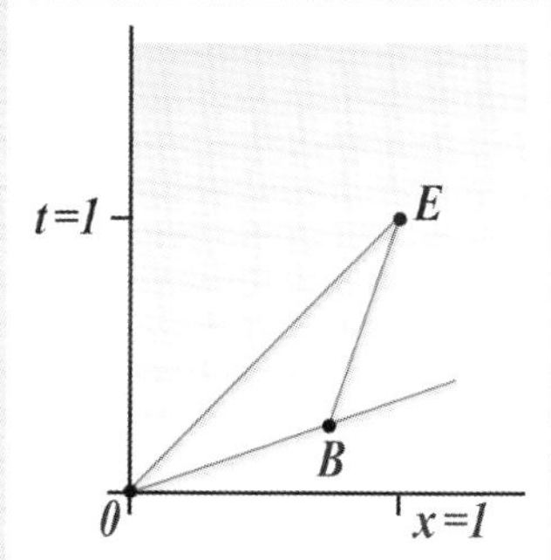

Fig. V: Mobile frame of reference

But you see the triangle 0-E-B has two identical angles, so these sides are also the same length and distance/time = 1 in both cases.

And then there was the most famous formula in the world:

$$E = mc^2.$$

I wanted to see how the basic assumptions had led to it, but the results of my inquiries were disappointing. What literature I found was by far not as direct as the three effects, but observations of special cases which did not deal with the main application at all. These were discussions of the possibility of a hydrogen bomb.

In the summer of 1945, my father received the most exact measurements of the masses of light nuclei from Joseph Mattauch, a colleague of his. Using the formula above, my father calculated the massive amounts of energy that would be released by nuclear fusion. Therefore it was important to know if these kinds of considerations fell within the scope of validity of this superformula. The question was if the different forms of energy of the components of atomic nuclei could actually change their inert total mass the way the formula demands. In the mass spectrometer from Mattauch, the nuclei are accelerated through an electric field and in this way their inert masses can be measured using the Newtonian law:

force = (inert)mass × acceleration, or mass = force / acceleration

These can be used to determine the amount of stored energy they have inside. Despite the obvious importance of this question, I wasn't able to find any proof anywhere that the famous formula could be used here. Originally, Einstein argued for the plausibility that the kinetic energy contributes to the inert mass, but whether the potential energy does as well is not so clear. After all, the potential energy is derived from the nuclear forces which have nothing to do with the electric force used to determine the inert mass.

How can it be that without knowing the exact nuclear force, I can postulate that the potential energy created hangs like a dead weight around the neck of the particle when I try to accelerate it? Looking back today, I see that in the case of two bodies this was only explained in the equation Hans Bethe and Edwin Salpeter made several years later. The fact that the nuclear powers behave exactly this way follows from the assumption that they also comply with the theory of relativity.

Back then, to at least arrive at a preliminary result, I worked out an equation for the two body problem in quantum mechanics that could only approximate the requirements of the theory of relativity, and therefore could only approximate the changes in the inert mass of the whole formation. This work wasn't a great success, but at least it gave me the feeling of having gotten my teeth into a fundamental question. Only after I had delved deeper into the details of the theory was I able to recognize the extent of Einstein's visionary mental powers.

The crucial step was his postulate that

(i) nothing can move faster than the speed of light,
(ii) systems moving uniformly in relation to each another are physically equivalent; in particular the speed of light is the same in all of these systems.

At first, this is counter-intuitive. Intuition says that when we run towards a ray of light, it would have to appear faster than when we run away from it. But if this were true, it would have catastrophic consequences for physics. We're always running around in different directions. With the earth's surface we run around the earth once every 24 hours, around the sun in a year, and with the sun around the Milky Way. If we measure something where the speed of light appears, it would have to be different in the morning and in the evening, in the summer and in the winter. What confusion! The highly precise measurements of today, often exact down to a billionth of a second, show us that this doesn't happen. Einstein was therefore right when he postulated that the laws of nature are the same in all systems which moved uniformly in relation to one another.

What should we do with our intuition telling us that the speed of light should vary? Our drawings have shown us how we can correct our thinking. To determine the speed of light, I need to divide the distance traveled by the time elapsed. These are different in both experiments, but the difference is canceled out proportionally.

Dublin and Glasgow (1949–1950)

My journey out in the world did not go as smoothly as it would have today.

Now you can step into a jumbo jet and just as if on a magic carpet fly over all of the enchanted lands like Turkey, Persia and India and land in Thailand nine hours later.

This can't be compared to what it was like to travel back then. Trains were the only form of transportation, and they moved at a snail's pace because the tracks were still damaged from the war. At each border there were hour long waits and passport controls, while every passenger dug out their passport and visa. This meant that even trivial distances, for example between Vienna and London, would mean sitting around in the train for several days and nights. I personally had to deal with these additional problems.

Problem 1: I was not yet internationally presentable as I spoke absolutely no foreign languages. The bit of English I had in school was useless for all practical purposes and German was neither understood nor appreciated abroad.

Problem 2 : Austrians were not allowed to either possess or exchange foreign currency.

Problem 3: The various bureaucratic procedures were not yet coordinated with each other. I could receive a visa for Ireland when I could show that I had an English transit visa which I could only receive when I had a visa for Ireland.

While I could do something about Problem 1, I was powerless in the face of Problems 2 and 3. Luckily a friend of mine, Verena Lieben, had married Heinz Laube, a diplomat who had accepted a post working at the Austrian embassy in London. He was able to help me. He lent me ten pounds and sent me an invitation to London which enabled me to get an English visa. In London, he was able to arrange for the Irish visa. I was finally able to set out in the Irish Sea and to arrive in Dublin. I can no longer recall exactly how I found Schrödinger there, but however it was, it all worked out for me.

Fig. 9: Erwin Schrödinger

Schrödinger lived modestly in Clontarf, a small suburb of Dublin, and he very graciously invited me to stay with him at the beginning. He didn't even allow me to help around the house, not even to dry the dishes. Only on one occasion did I have to get on my bike and buy some tobacco for him on a Saturday, because he couldn't bear the idea of a weekend without cigarettes. It didn't seem to bother him that his chain smoking would ruin his health by the age of 70.

There was a relaxed atmosphere at the Dublin Institute for Advanced Studies and we could do as we pleased. This doesn't mean that nothing got done, on the contrary; everyone worked as much as they could, and had to determine for themselves where their breaking point was. (Unfortunately, not everyone was able to do so.) Schrödinger would come to tea late in the morning and discuss everything under the sun. He preferred to criticize other people's publications rather than report on what he was working on at the time, which was a mechanism through which particles of an electromagnetic field could receive a mass.

Even we didn't know this, and the people who weren't in his inner circle really had no idea. Only 20 years later was this rediscovered by Peter Higgs and others, and is today the credo of particle physics. This lack of communication let one of the greatest discoveries of the 20th century, the non abelian gauge theories, pass right by the Institute in Dublin. They had all of the ingredients. Schrödinger was working on field theories in which the so-called connection plays the main role. In the next room, Walter Heitler had found that the isospin, the rotations in three-dimensional space, is built on the group O(3). These two things taken together produced a theory that the great Chinese physicist Chen Ning Yang formulated and is today the prototype for all elementary particle theories. If Schrödinger and Heitler had been more interested in the physics of each other, this theory would have appeared ten years earlier.

Schrödinger was also hoping to find something like a "theory of everything" and he came up with the right acronym GUT (Grand Unitary Theory, today U is read as unified) but that was all.

He was not only the scientific star of the Institute, but also its director. He was not the ideal candidate for more administrative positions as he was too highly strung. A French colleague once told me that the atmosphere was strained when she went to visit the Institute. Both Schrödinger and the cleaning lady had handed in their notice. The point of contention was the fact that she had emptied his wastepaper basket in the evening, but he had thrown away something that he still needed. He complained that she was sabotaging his scientific work, and she complained that he was preventing her from fulfilling her duties.

Schrödinger had a perfect command of several languages. He wrote poetry in German and English and his scientific books are also a literary delight. I had the pleasure of hearing his lectures which he used as the basis for his book Space and Time Structure. It was a strict reading of the manuscript, not a free exposition. This worked fine as long as he didn't come across a formula. Then he had to put away the manuscript and go write it on the chalkboard. He often got lost in going back and forth. From the seats in the audience, where there was an overview of everything, it was easy to give pointers to get back on track. Some people are irritated when others point out their mistakes. Schrödinger wasn't, he was truly grateful.

One captive member of the audience was the mathematician John Syngh. He noticed that even the simplest desiderata like proofs of existence or uniqueness of the solutions were missing. This was not due to carelessness on Schrödinger's part, but was entirely consistent with the state of understanding at the time. These kinds of questions were not part of the zeitgeist. Syngh uttered the prophetic words that it would take another half a century until these mathematic gaps would be filled. He was right.

At Glasgow University

Fig. 10: Bruno Touschek

Bruno Touschek never became as famous as the other people mentioned here. He was more like a boy scout and never really grabbed the spotlight for himself. There are still many legends about him, and in Italy there is even a 90 minute movie about his life. He wasn't a fan of formal occasions. I once had the opportunity to experience him letting loose when a conference at Lake Balaton culminated in a wild party.

In Glasgow, he was still leading a very down-to-earth life and had no vices other than smoking. In our spare time, we went to the theater or played badminton, but much more aggressively than children do now. Later in Rome, he developed two more hobbies, wine and motorcycles. The two don't go together very well, and they brought him into conflict with the authorities. A police officer noticed his accent and asked him what he was doing in Italy. Bruno answered truthfully that he was working. The officer wanted to know what kind of work he was doing. Thus, Bruno continued telling the truth, but it would have been better if he hadn't. He was working on the question of whether the basic laws of physics determine the direction of time; if the time reversal of a motion would be an allowed motion. Bruno answered the officer that he was working on going back in time. The officer decided that the man was not only drunk, but also insane, and Bruno was taken to a psychiatric clinic.

He often uttered the wrong thing at the wrong time, even during his wedding. He had been able to resist the charms of Scottish women while in Scotland, but ended up falling in love with a Scot in Rome. The wedding took place in Scotland where it was not merely a simple matter of saying "I do". A detailed statement must be made which includes the sentence "I shall always be faithful to my lawful wife". With all the stress Bruno managed just to say

"I shall always be faithful to my awful wife".

In physics, he was always aiming high and was a major influence in the development of particle physics. He was interested in joining the fundamental laws, but realized that this could only be possible with greater energies than could be created at that time. You would still need about 100 times as much energy, and that just didn't seem possible. If you shoot a particle with greater energy on an inert particle, the latter yields, and in so doing consumes energy. The remaining energy only increases with the root of the primary energy. To reach a factor of 100 for the remaining energy when measured in proton masses, the primary energy would have to be increased 10,000 times. This cost is simply too high.

The only alternative was to shoot two energetic rays directly at each other, but this seemed like a pipe dream. These kinds of particle rays are extremely thin and are not much more than empty space. A solid piece of matter at practically the speed of light would be a dangerous thing, worse than a nuclear bomb. Because the rays are mainly empty space, it would be very rare for two particles to run into each other, just like when two galaxies come together only a couple of stars at most collide.

It took a long time to hone the rays, which were at first as thick as fingers, down to be as sharp as razor blades and to then aim them at each other in a way such that a significant rate of collision was generated.

Touschek was the driving force behind this development. He left Glasgow for Rome, because there were plans to build such facilities in Frascati outside of Rome. First a smaller prototype called Ada was developed, and after it had been proven that the principle worked, a larger model called Adone followed.

This developmental work takes time, and by the time of Adone, it was already the turbulent year of 1968. Italy was particularly infected with strike fever and this held up the completion of Adone. In the meantime, this technical advance had been admired and then imitated around the world. A laboratory in Siberia was even able to produce the first physical results, beating out the Romans by a nose.

This was naturally devastating for Bruno's hypersensitive nerves, and the empty wine bottles began to stack up in his apartment. It took a long time for the technique of the colliding beams to find their way into, and ultimately dominate, high energy physics. Like Moses, Bruno could only see the Promised Land from afar, but could not enter; his cirrhosis was too far advanced. I spoke with him on the telephone on the day he died. He said that his life had no meaning anymore. He dismissed my objection that the times would change. But they did. The colliding beams are the only way to throw back the gates to a world that had previously seemed unattainable.

In Glasgow I was able to present what I had developed in Dublin, a clearly stated derivation of the fundamental formulas of quantum electrodynamics using what was at that time the modern formalism of Freeman Dyson, Richard Feynman and Julian Schwinger.

To learn this method, I thought it would be best to try it out on something already known. This worked perfectly, and so I wanted to share it with others. It suited me fine that I also had to give lectures as part of my job in Glasgow. In Göttingen, I wanted to publish this material as a small booklet. The Springer publishing house had an office in Göttingen, and I offered them my manuscript. Springer at first seemed interested, but ultimately rejected the final manuscript. So I filed it away where the material stayed for some years, until one day when I happened to walk by a second hand bookstore in Vienna which also happened to run a scientific publishing house, "Franz Deutike". I went in without any great expectations, but to my surprise the manager was interested in it. He just didn't like the title "Quantum Electrodynamics". He said that it was mostly textbooks that were now in demand in the scientific sector. Mine just wasn't a textbook in the usual sense, so we finally agreed on "Introduction to Quantum Electrodynamics".

To prevent it from being a complete misnomer, I included a purely intuitive estimate of the magnitude of the different effects. This gave the booklet a special character and it was soon sold out. It was first translated into English and then into Russian. The Russian translation became a real bestseller with 50,000 copies sold. When I asked how this could be possible, I received the disillusioning reply, "Well, here all scientific books are cheap because the state subsidizes them. In the far corners of Siberia paper is expensive. That's why farmers buy these kinds of books and use them to roll their cigarettes".

The time in Ireland and Scotland was a complete change of scenery in my life. Ireland was a country which had not been at war, with a seemingly easygoing population, the powerful sea, the soft hills inland; it was a whole new world. In Austria, we were still living in rubble heaps. Nevertheless, the Alps were more magnificent as mountains, and I carried with me an unforgettable impression of their majesty. Towards the end of my time at university I had gone mountain climbing with a friend and two companions in the Dachstein area. We had already reached several summits and one clear afternoon we saw a splendid rock face. We should be able to climb that and return by evening, we thought. However, we had underestimated the rock face and overestimated our abilities. One of the women turned back quite soon. Night fell and the three of us were still on the cliff. Our friend, who was a little older and a better mountain climber than I, had the situation under control. He showed us a small ledge and said we should sit and wait there for the morning. He would continue up and go down the easy way. Early the next morning he would come get us out with a rope. That's what happened. We spent the night huddled together, forcing ourselves not to fall asleep as dozing off would have been fatal. My companion was a brave woman and never complained about the dangerous situation we had gotten her into. We were lucky, there was no sudden fall in temperature, as the last thing we needed was an ice covered cliff. But then something wonderful happened, the moon rose slowly and transformed the glacier far below into a glowing ledge.

We couldn't have thought of a more beautiful way to spend our time until morning than gazing at this amazing scenery.

My memories of Ireland are more peaceful and less dramatic. The moon also plays a role here, the way it turns the moorland golden. I continued to go hiking with my colleagues from the Institute on the weekends, but hiking in the Wicklow Mountains is just not the same as mountain climbing in the Alps. Hiking in Ireland does have its appeal, it's pure relaxation without any physical challenge. There are no crowds, and it's more or less deserted inland. We set off one afternoon and thought to ourselves we'd find somewhere to stay somehow. But there was nothing other than gently rolling hills covered with thick brush, overgrown with heather. Once again in my carelessness I had exposed myself to the night, and the moon was my sleeping partner. This time, though, the heather made a soft pillow and I could find comfort in the lap of Morpheus.

All in all my time in Ireland was idyllic, even though there were traces of poverty that could not be overlooked in Dublin. There were rotting houses with droves of dirty, ragged children playing in front of them. At the same time, the country allowed itself the luxury of the Institute I was working for. It simply enjoyed the government's goodwill. Once when I later visited as a professor, my wife and I were even invited by Ireland's founding President de Valera to his private residence. This investment in education paid off for the Irish when they joined the European Union. They knew how to seize the opportunity and soon they had a higher per capita GDP than in Great Britain. In my time, they were still suffering the stigma of being the region's poorhouse. Another thing I had thought impossible then began just 15 years later. Behind a religious-ethnic cover, fanatics corroded the society and began to blow each other up. I was not able to tell if someone was Catholic or Protestant or if they came from the North or the South. I have no idea how they can even tell who they want to kill. I am appalled that something like this can happen in Europe today.

Although I had formed a connection to Ireland and its people, I had to continue on my way. This would lead me to Werner Heisenberg in Göttingen and Wolfgang Pauli in Zürich, where I could share experiences with other creators of quantum theory.

Göttingen and Zürich (1950–1952)

Fig. 11: Werner Heisenberg with Walter Thirring

Göttingen was a small town, but one of the scientific capitals of the world. The name of Heisenberg held a magical kind of attraction for up and coming physicists, and that's where I met the people who would spread theoretical physics throughout Central Europe. Harry Lehmann, Kurt Symanzig, Wolfhart Zimmermann, Reinhart Oehme and Gerhart Lüders had all gathered there from Germany. There were also many from abroad, Jurko Glaser from Zagreb, Paolo Budini from Trieste and Bruno Zumino from Rome. Unfortunately, Heisenberg was so busy with other things as the director of the Max Planck Institute that he could not devote so much time to us. That didn't matter so much because we had so many other sounding boards available. Once a week we had a group colloquium, which Heisenberg always made a point of attending. We were able to learn a lot from his way of thinking, especially as he tended to do his thinking out loud. For example, I was there when Heisenberg introduced spontaneous symmetry breaking. It was during a lecture on superconductivity.

Lecturer: ...and that shows that there is a current flowing in this model in its ground state.

A critic: But there's no direction given in your model, how does the current know where to go?

Lecturer: It just flows somewhere.

Critic: And why not in the opposite direction?

Heisenberg: It's probably like ferromagnetism where the individual domains set themselves up somehow. There are small random disturbances which determine the direction.

Critic: But I don't see any disturbances in this model.

Heisenberg: But they're there in reality, and that's what's important.

Critic: I thought that the lecturer was talking about a mathematical deduction.

Heisenberg: We're here to do physics, not math.

In Göttingen, there was also an active intellectual life outside of the Max Planck Institute. There was a good theater, and I also could enjoy plenty of music, as there was a marvelous Blüthner grand piano in the Nansen House where I was staying. There was also a physics department at the university, but strangely enough there was hardly any contact between the department and the Max Planck Institute. During the entire year I was there, not a single person from our group was invited to give a lecture at the university. It certainly wasn't due to any incompetence on our part, but perhaps because many people at universities were suspicious of the physics Heisenberg was doing at that time.

The quantum field theory was experiencing a Renaissance and we were studying it with gusto. But Heisenberg wanted more; he wanted to be able to put an "end" to physics. He thought that there would be nothing new to find if even greater amounts of energy were used. We already had everything we needed and, according to him, now just had to pour it all into a "theory of everything". He wanted to experience this triumph in his lifetime, and he wanted the theory to come from Germany.

The last decades of his life were consumed by this tenacious task. He had no use for anything that might call this simple idea into question. However, he was beating his head against a wall he would never be able to break through. The fundamental protagonists are only found above the energy boundary surrounding his world. His feeling that there would be a unification of elementary particle theory has proven to be correct in the standard model of today. His dogma that all particles can be constructed from spin ½, or as he put it, "everything spins half" is only partly true. Besides quarks and leptons there are also gauge particles. Supersymmetry says "half spins half". Heisenberg's impression of the reach of elementary particle theory was greater than what most physicists expected. This could be seen at a conference where Heisenberg presented his usual lecture and Gregor Wentzel was sitting in the audience. The weakness in Heisenberg's theory was that he did not want to acknowledge the newly discovered particles. This meant that not all of the baryons were included in his theory, which led Wentzel to ask afterwards if his theory could explain everything about the physics of baryons. However there was a lot of background noise in the auditorium and Heisenberg didn't understand the question correctly. He briefly looked up at the ceiling for inspiration and then said in a strong voice full of conviction, "Yes, biophysics is completely covered in the equation". What a theory of everything!

Heisenberg has often been criticized for his behavior under the Nazis, unfairly in my opinion or at least not with enough distinction being made. Of course he was too intelligent to fall for Hitler's primitive slogans, but he was "nationally minded" as they say. He wanted to serve his country and had always refused offers to work abroad. Sometimes his patriotism went too far. Personally I never heard him say anything objectionable, but Bruno Touschek once told me about a time when all of the civilian casualties of a military action were mentioned and that Heisenberg had replied, "but they were just Poles".

For Heisenberg, the Nazis were just a political party like any other, and he didn't appear to be too horrified by their crimes. Naturally, the Nazis also used his name for their prestige and he even became the director of the German atomic program. The fact is that the Germans never even came close to the atom bomb, mainly because Heisenberg had said it couldn't be done. In a report he had listed the critical mass, the mass needed for it to explode, as over 100 kg, which is actually ten times the amount necessary. I can't say if this was done on purpose or not. I once heard Carl Friedrich von Weizsäcker's answer to the question of whether they would have built the atom bomb for Hitler if they could have. He said that he was no longer able to put himself in that situation to be able to answer truthfully. In any case, German nuclear physicists were successful in protesting against Germany becoming a nuclear power after the war.

Many things will remain a mystery for modern readers, just like I can't understand some things in Andre Sacharov's autobiography. All Russian physicists took part in developing the hydrogen bomb for Stalin, even though they knew that Stalin placed no value on human life.

The only physicist who preached against building an atomic bomb was Walter Heitler. He had been a student of Sommerfeld, and as a Jew he first had to emigrate to England, then he went to the Institute in Dublin and finally, after the war, he became a professor at the University of Zurich. I got to know his opinion about the atomic bomb in the early 1950's when the hydrogen bomb was developed. He did not tire of repeating in schools, churches, wherever, what a horrible crime it was to cremate 100,000 innocent people in a single second. This was why he strictly rejected nuclear armament. Unfortunately, his followers carried no weight in the media, and he was hardly taken notice of by the public. The situation was very different for his opponent Eduard Teller, who was a proponent of building the hydrogen bomb, and then for nuclear armament. He was always making headlines in the papers, and then he was also showered with peace prizes. Today, we're stuck with around 100,000 nuclear warheads; we will have to spend inordinate amounts of money for their disposal and also have to worry that they might fall into the wrong hands.

The next station led me to Wolfgang Pauli. I had the good fortune to be his scientific assistant at the ETH (Swiss Federal Institute of Technology) in Zurich from the fall of 1951 to the summer of 1952. At twenty-four, I had already published a few scientific articles, but compared to the great Pauli, I was a complete nobody. He never made me feel this way, but always treated me as his equal. This doesn't mean that he was all sweetness and light. Once, for example, he said to a colleague, "Mr. X, I don't mind that it takes you so long to think, I just mind that you publish faster than you think".

Pauli left it up to me which problems I wanted to solve, and I wasn't exactly modest in my choice. At that time, Pauli and Heisenberg had paved the way for a development to become a certainty, and it would transform our understanding of the nature of matter.

The Essence of Matter

The classic picture of matter breaks it down into its permanent parts, the atoms. These are something like little balls that flit about, but that's all. In between there's just empty space, or nothing. In the new picture, the primary building block is the field. This is the carrier of all possible materials and fills all space. What we consider to be particles of matter are just local energy stimuli of this field. They are by no means permanent, but rather are subject to a constant emergence and disappearance. They make up what we can call the visible world.

The unstimulated part of the field builds a kind of world of shadows let's call the underworld. It fills all space even where there are no particles and contains the plan for all atomic particles. These do not exist in the underworld in actual form, but as a possibility. These potential particles have been haunting the media recently as "dark energy". The figures given below will prove that we have precise knowledge about them. When enough energy is invested, then particles can be transported out of the underworld into the visible world, just like when the α-particles break out of the atomic nucleus. Even if you're poor and can't afford a high energy machine, an occasional visitor from the underworld will stop by (virtual particles). For the electron these are photons, positrons and other electrons, which swirl around it about 1% of the time and shake its magnetic moment. This is then changed from 1 (in suitable units) to 1.001 (first generation).

These visitors do not show up alone, but have brought others with them, who change the value by 10^{-2} with the second generation determining the fourth decimal place. At that time there were precise measurements taken that confirmed the effects of the first generation of visitors, and experimental and theoretical physicists have been spurring each other on ever since. The experimental physicists have been making more and more precise measurements which force the theoreticians to calculate the influence of further and further generations. This forces the experimenters to make even more precise measurements until today we have the magnetic moment μ of an electron as

$$\mu_{Exp} = 1.00115965219 \pm 0.00000000041$$
$$\mu_{Theor} = 1.00115965219 \qquad (2.1)$$

and have achieved the most exact agreement between theory and experiment in physics. The $\pm$ at μ_{Exp} shows the possible error margin.

The discerning reader will find it a bit suspicious that there's such a large degree of agreement between theory and experiment and will perhaps think that there's something fishy going on. Well, to a certain degree there is. μ_{Theor} is dependent on natural constants which cannot be exactly measured individually. The experimenter takes μ_{Exp} as a measurement of these constants and adjusts them so that μ_{Ex} and μ_{Theor} form a perfect match. Some of my colleagues find this approach to be a bit amateurish and prefer to use the average of other measurements for the constants. This changes the last decimal place slightly, but the level of agreement remains impressive. It shows how precisely the picture of the visitors from the underworld predicted by the theory was right, despite quantum mechanical fuzziness.

I wasn't so fascinated by the exact figure then; it seemed to me to be more of an amusing pastime which could occupy simpler minds. For me, it was the principle of the thing, the image of the visitors from the underworld. Would the effects of the following generations become smaller and smaller, or could they snowball and cause an avalanche? I had the feeling that something like the latter was possible and wanted to come up with the mathematical proof. The time and energy needed to calculate the magnetic moment of an electron was much too great, so I chose a simpler characteristic of a hypothetical particle that had the same mechanism. For simplicity's sake, I chose the probability for the diffusion of two particles at their point of collision.

I didn't need any fancy abstract terminology for the proof. I just had to count: the n-th generations will appear with the probability of α^n*, with the coupling constant* α *being something small; for example, in quantum electrodynamics it's 1/137.*

Therefore, I only corrected the effect in the first generation by 1‰, the next generation again less this factor, so that the procedure would seem to converge nicely. But it can add up when there are particles everywhere and all effects have the same sign. It turns out that the n-th generation in $n^{3n/2}$ *different constellations contribute to the effect, but some constellations are equivalent and should not be counted twice. The equivalents come about through permutations, and it's been determined that around* n^n *they are equivalent to each other.*

So $\alpha^n \cdot n^{3n/2}$ *should be divided by* n^n*, but then there's still* $(\alpha\sqrt{n})^n$ *left over, which at first only seems to grow smaller. As soon as n is greater than* α^2*, it increases so as to be incalculable.*

This was the somewhat naïve calculation that I had to keep in mind, the carrot on the stick. But then doubts arose. Is this really true for any size of n, do all terms have the same sign, are they as large as we think? Then if you find a mistake, can it be repaired or is it fatal?

While working out these ideas, I experienced a myriad of emotions, but I will stick to a telegram style in retelling these events. I can reconstruct the chronological order because at the time I wrote my fiancée (now my wife) twice a week and she kept the letters. For the curious reader, I've included the final letter about this topic as a P.P.P.S. My emotions were flying high when I wrote this letter describing my mathematical operations, and it was by no means all smooth sailing. Here is a condensed version of my highs and lows:

December 1, 1951: I'm agonizing over how to get to the root of further generations when it already takes months for the best calculators to come up with the second generation.
February 1, 1952: Trash can filled with mountains of calculations, but no real progress.
February 10, 1952: The effects of the tenth generation are already so tiny that I'm already far beyond what can ever be measured.
February 15, 1952: In the 50th generation the effects are still only 10^{-50}.
February 20, 1952: I'm able to prove that, even if I can't precisely calculate the effects of the nth generation, this number must at least be larger than a positive number specified by me.
February 21, 1952: Aha! At around the 60th generation, the diminishment of their effects begins to slow down.
February 22, 1952: Pauli enters the room, "Mr. Thirring, the ETH is going skiing next weekend, wouldn't you like to come?"
Me: "That sounds nice, Professor Pauli, but I'd rather finish my work here".
Pauli: "You can put together one of these little commentaries that you're always publishing in the ski hut".
February 23, 1952: The trend continues, and starting with the three hundredth generation the effects are as great as in the beginning. I really did set off an avalanche.
March 1, 1952: I proudly report my findings to Pauli, who says, "But how did you make the individual results finite, when they're in fact infinite?" I answer, "I used the regularization that you published with Villars". Pauli: "That was just a mathematical trick we did. Today it's known that this is the difference of large numbers, and if your results are to have any meaning in physics, you will need to estimate this difference". (In modern jargon, renormalization instead of regularization.)
March 2, 1952: I realize that the limits for this difference will be an even more difficult problem, but after what I'd been through, I'm fearless.

April 1, 1952: Trash can filled with mountains of calculations, but no real progress.

May 1, 1952: I'm able to show that this difference isn't much smaller than my first result, which can therefore still stand.

May 2, 1952: Pauli enters the room, "You've gotten some competition. I'm getting an article from Angas Hurst, an Australian who's now working in England. He has made the same calculations and has arrived at the same conclusion. Why don't you take a look?"

May 3, 1952: Hurst's article contained my first result, but not my others. Thanks to Pauli's criticism, I'm ahead by a nose. I can publish my article, and it doesn't have to join the mountains of calculations in the trash can.

Of course my calculation had to be examined more closely, and Gunnar Källen was, in fact, able to find a mistake. But this could be rectified and the result still stands.

My colleagues reacted thus:

Richard Feynman: I don't give a shit whether it converges!

Bruno Feretti: Most functions are non-analytic.

Markus Fierz: There's no way it could converge because there's no limit to the energy level going down in this theory.

Gunnar Källen: I don't like Fierz's objection because through the normal order no theory has positive local energy.

Freeman Dyson: If the series were to converge, it would do so even if the electric charge would be completely imaginary. But then the charges with the same signs would be attracted to each other and the state of lowest energy would be a clump with all of the positive charges and one with all the negatives. Their energy would go faster than the number of charges against $-\infty$, so that energetically it would be better to constantly create new particle pairs and never have stable conditions.

Heide Narnhofer and I substantiated Dyson's argument mathematically the other way around. If you don't let particles of any sign clump together, then this series will converge.

***

P.S. Angas and I later became good friends, and our articles — together with one from André Petermann, who independently came to the same conclusion at the same time — are still present in the literature.

P.P.S.: Even today, a half a century later, it's still not clear why electrons and photons don't have these avalanches of visitors. The fact that they could has been proven in numerous experiments and is undisputed. The creation of the universe, the big bang, was probably such an avalanche. Its particle flood was made up of virtual particles out of the underworld which had been set free by gravitational energy.

P.P.P.S.
My letter from May18, 1952:

My dearest Helga,
It's over, I've scaled the last peak. I have finally solved the problem that all of the educated gentlemen have been fooling around with over the last five years and none of them have been able to put together. You can imagine that it wasn't easy. The people who wrote the long calculations that I showed you aren't easily intimidated by a few integrals. Please excuse the fact that my last few letters have a bit more meager than they should have been. Pauli is very excited and like I've already told you, he managed to secure enough money for me to be able to stay until the summer. I also have to go to Copenhagen where there will be a big physics congress at the beginning of June. I'll have to give a lecture there because my work now is pretty much the most important thing to have happened in theoretical physics recently. You know, I don't really care about becoming famous, but I've been blaming myself ever since I published that nonsense with the wrong signs, and am therefore so happy that I've been able to solve this problem to make up for it. I can briefly tell you how everything came about: about

four weeks ago I read something about this problem and the man claimed that this problem was unsolvable. You know how something like that rouses me to prove them wrong. And the problem does seem insurmountable from afar, but when you just go and take the bull by the horns, it's not so bad by half. I was soon able to make considerable progress towards the solution, and the calculations were also not extraordinarily complicated, but extremely dangerous. A calculation is always easy when you know what the answer has to be, because then you can steer yourself in the right direction. But when you want to explore uncharted territory, there's always the danger of falling into a trap and landing in a dead end. Finally, exactly on my birthday, I was able to put the finishing touches on my chain of proof, and everything seemed watertight. Pauli was very excited, and I set about writing everything down. You can imagine how I felt when I noticed that I had gotten caught in a trap and that one of my arguments was completely wrong. I had to go back once again and find a new way. That was the moment of real despair in this whole time, because it wasn't certain that a new way even existed, and on the other hand I had to find one, otherwise Pauli would think that everything I'd said was all made up. Every day for two weeks I'd think I'd found a way to salvage everything, only then at night to discover where the catch was. It was like I was Penelope, each night undoing what she'd woven during the day. It often occurred to me that what I was trying to prove wasn't right anyway. The saving grace came when Schafroth remembered that the area my path was taking had been explored by one of the great mathematicians of the century. I had actually already given up on this path, but didn't want to leave any stone unturned. In fact, it turned out that the results of the mathematician deliver exactly what I'd been looking for. Now that nothing has changed in the last four days and I've checked everything from the very beginning several times, I think I can be sure that now everything is correct.

***

That same year Pauli made it possible for me to present my results at the International Conference of Theoretical Physics in Copenhagen. At that time Copenhagen was the Mecca of physics and it was a great experience for me to feel how Niels Bohr dominated the scene without making a specific scientific contribution.

But a new era ushers in a new spirit, and at that time it was to be found in the young American Arthur Wightman. He had received a modern education in mathematics at Princeton and invoked general mathematical theorems in the discussions. This annoyed the math dilettantes in the audience, and provoked many different reactions. That was when non-local quantum field theories were emerging, but no one could find a Hamilton function where the equations could come from. Pauli had a private consultation with Leon Rosenfeld to determine the official party line on this critical question. Finally, they announced that these are theories that are Lorentz invariant, but there is no Hamilton function. There was a general feeling of relief, but then a tall, wiry figure spoke up from the back:

Wightman: That's not possible, that contradicts Stone's theorem!

Pauli's countenance darkened and he just managed to utter, "Why should that contradict anything?"

Wightman: "Lorentz invariance means that there is a unitary representation of the Poincare group and Stone's theorem says that its elements are created by self adjoint operators. Time is generated by the Hamilton function and therefore must exist in every Lorentz invariance theory".

It was a blow to Pauli to be so coldly dispatched, but he remained calm and consulted with Rosenfeld. He was an expert in many areas, but modern mathematics was not one of them. Neither of them were familiar with Stone's theorem, and therefore did not know its conditions. If they had, they could have easily given Wightman enough rope to hang himself. They could have dismissed his objections with the argument that the continuity requirement of the theorem had not been fulfilled. This wouldn't even have been so impossible; we were later easily able to provide similar examples. But Pauli didn't know this, and it wasn't his way to just make something up. So he declared that the entire thing was heresy: "You must take back this argument, this is unacceptable".

Wightman did not budge an inch. "This is one of the fundamental theorems and no one can doubt its validity".

Pauli was impressed by so much relentlessness and he simply said that this had to be explained better.

Wightman was able to do so the next day, and Pauli had to concede that he was right.

Pauli's personality has been described as having two main characteristics, a somewhat rough way of dealing with people and the so-called Pauli effect. I did not suffer any lasting psychological damage from the first, but I remember well the Pauli effect. This was named after his ability to bring out the weak spot of an object, and break equipment or have things go wrong simply by being in the vicinity. Experimental physicists were understandably nervous when he came around, but other things could happen as well. The following is a typical legend: Wilhelm Backhaus was giving a piano recital in Zürich, and as Pauli enjoyed going to concerts, his assistant bought tickets. Everyone familiar with the effect was waiting to see what would happen, if Pauli would make Backhaus hit the wrong note? Nothing of the kind happened, Backhaus and the Beethoven sonatas were as solid as granite, nothing could be shaken. But then everything turned out to have gone wrong anyway, because Backhaus played the wrong sonata.

In 1955, I witnessed the effect, and saw how Pauli enjoyed the completely irrationality of it. We had organized a conference in Berne to honor Einstein, who was ultimately not able to come personally, but Pauli came and was the center of attention. The highlight of the conference was a lecture by the astronomer Baade, who spoke about cosmic expansion. Naturally, we had taken all necessary precautions. The lectures took place in the municipal museum in Berne and this was filled with Swiss solidity. Just to be on the safe side, I asked if there were spare lamps for the projector, and received the indignant reply that the projector had been working for the last 80 years and there had never been any cause for complaint. I let it pass, but was a bit concerned when I saw that Pauli had taken a seat next to the projector for the main lecture. What could I do? At first, everything was fine and the lecture proceeded to come to its climax. It was a picture that would unmistakably show cosmic expansion. But nothing came of it, there was just a flash of light and then absolute darkness. The eeriness of the scene was heightened when suddenly Pauli's sardonic laughter rang out, breaking the silence.

Berne (1952–1953)

Fig. 12: Fritz Houtermans

Once again my life underwent a complete change of scene — Berne instead of Zurich, Houtermans instead of Pauli, a position as faculty member at the university instead of a fellowship, and most importantly, reaching the haven of matrimony instead of bachelorhood. After an unhappy first love which might even have been the cause of my weak scientific achievements during my time in Göttingen, I finally found the one. She has been one of the best things ever to happen to me. She was one of the luckiest strokes of fate in my life. She brought stability to my private life and our marriage has lasted until today.

Berne has a completely different character than Zurich. Zurich is international and cosmopolitan while Berne, although it's the capital, is provincial. It was very small and cozy, not stressful, and was good for our young marriage.

Fritz Houtermans was not only different compared to Pauli, but different from every other physicist I knew. This was probably due to his terrible fate. He spent his childhood in Vienna and then went to Germany to study and work until Hitler's rise to power put a stop to it. He was considered only a "half Jew", but had expressed himself so far to the left politically that he wanted to avoid the Nazis at all costs. So he emigrated to England, although he wasn't able to find a job at any university, but only an industrial job in the provinces.

The industry did not know what to do with him and his talents — he was, after all, the first person to understand how lasers work. He decided to accept an offer to go to Kharkov. The USSR was going through a liberal period and foreigners could be invited to teach at the universities. At the time Lew Landau, the leading Russian physicist, was working in Kharkov, and he held an immense attraction. This was therefore an offer Houtermans could not refuse, and at first he enjoyed this change.

But then Stalin began looking for scapegoats, and entire ethnic or professional groups like railroad employees or peasants were taken to the gulags. Foreigners in particular were ostracized and despite international protest, Houtermans was arrested. He was interrogated and ultimately tortured to finally confess to being a spy. Thereupon he was also sent to a gulag. Only after the Hitler–Stalin pact were the Germans released.

For Houtermans, this release did not mean freedom, because for the Nazis, of course, he wasn't a German spy, but a tried and true communist. So he was thrown in jail again, this time in a Gestapo jail.

Luckily Nobel Prize winner Max von Laue heard of poor Houtermans' fate, and tried to get him released. It was actually possible for von Laue to convince the German authorities that the German uranium project was doomed to fail without Houtermans, as it was Houtermans who had apparently demonstrated the possibility of a plutonium bomb. In fact, Houtermans was released and allowed to work on war related projects.

The long years in jail did not fail to leave their mark on his nervous system, and he was heavily addicted to coffee, whisky and cigarettes. Keeping in character, his first scientific achievement consisted of a postulation that tobacco would make a particularly good neutron moderator and he would therefore require a truckload of tobacco waste for his research.

This request was easily fulfilled as the waste would have had to have been disposed of anyway, and now Houtermans was able to choose from it what he wanted until the end of the war. It was also not such a tragedy that the neutron moderator didn't work out; a German atom bomb was in any case in the distant future. Once, when I asked him years later what he did after the war, he answered, "Then I bought myself a pipe". I asked, "Where did you get the tobacco?" He replied that he wandered around with it at night looking for English soldiers who were smoking cigarettes. He would ask them for a light and held out his empty pipe. When they put their cigarettes in the bowl, he would inhale deeply several times and was able to get a few drags from their cigarettes. He held on to his resourcefulness and creativity and had not become bitter despite everything; on the contrary, he was up for just about anything. I can't say that he was completely uncomplicated as a boss, but he remained benevolent, even when he was in a bad mood.

His scientific achievements have been largely forgotten, even in Berne, where he taught as a professor from 1952 until his death in 1965. Just recently I asked a visitor from Berne if people were still familiar with the name Houtermans. Our visitor shook his head and said that the oldest physicist he knew of was Leutwiler. Leutwiler had been a student of mine, so that would be two generations after Houtermans.

I would like to take the opportunity here to honor him, although I have to hesitate on the question of whether to call him an experimental or a theoretical physicist. Maybe we should agree on the term "explorer of nature". He took part in numerous international experimental projects; for example, in the measuring of the neutrons in cosmic radiation. Initiated by the University of Chicago, it was to take place at the Jungfraujoch ("virgin mountain") research station. Many other countries participated and they each bragged about the ingenious equipment they had. I will never forget Houtermans answer to the question of what Switzerland was providing; he simply replied, "the virgin (die Jungfrau)".

As for experiments in my seven years in Berne, I witnessed endless measurements of the rate of decay for radioactive samples. This was for rock dating and was connected to one of Houtermans' groundbreaking achievements. He is the first one to have ever determined the earth to be four and a half billion years old, a figure that still stands today. Houtermans then wanted more, he wanted to determine the age of the universe

At the end of the 19th century, Lord Kelvin had estimated the energy loss of cosmic bodies to see how long the energy supply lasted and came up with just a few million years. This is how long it would take to get rid of the collected gravitational energy of the sunshine. Such a period of time was much too short as the geological and biological facts required billions, and not millions, of years. This appeared to be a serious problem and all of Darwin's theory of evolution was called into question. And then in 1929 the 26 year old Fritz Houtermans together with his friend Atkinson had the idea of a different energy source that was a thousand times more powerful — nuclear fusion.

Of course, quantum theory was brand new back then and a theory of atomic nuclei didn't exist. It would still take a long time until a solid theoretical description of these events was made, but the time did come and showed that they had been pointing in the right direction.

Overlooking the lack of theory, there was also at first no experiment available to support this pipe dream of such a super energy source, because, of course, no one could look into the sun's center where this nuclear fusion should be taking place. Or could they? So he came to me one day and said, "You're so good at figures, can't you maybe figure out if it's possible to prove anything about the neutrinos produced during nuclear fusion?"

At that time, the neutrino still hadn't been discovered, and no one was thinking about the parity violating v-a interaction. I worked on the calculations as well as I could, but my results showed that an experimental neutrino physics was impossible. We could see that the scattering cross-section greatly increased with the energy, but the idea that one day a machine would be built producing this kind of energy, and that this machine would produce accelerated particles at a much greater intensity, was something straight out of science fiction. The experiments today where neutrinos — produced in CERN — fly through the earth to Gran Sasso where they are examined in a giant experiment hall underground were completely unimaginable, just like it seemed absolutely impossible to provide evidence of neutrinos on earth produced in the sun. Although we receive many of these, a hundred billion a second in cm^2, their impact is low as they are so tiny, with a diameter of 10^{-22} cm, so they can slip through everywhere. But sometimes one will hit a waiting detector in the middle of an atom, and even the direction from whence it came will be registered. So it's not only possible now to prove that there are neutrinos in nuclear fusion, but also that they come from the direction of the sun.

Unfortunately, Fritz Houtermans did not live to see this triumphant confirmation of his prophesy.

Princeton (1953–1954)

Fig. 13: The Institute of Advanced Studies in Princeton

The results I worked out and described in a telegram-style above got me an invitation to the Institute of Advanced Studies in Princeton, New Jersey. Naturally, I was excited to accept not only because Albert Einstein was working there, but also because the USA had become the leading scientific nation since the end of the last war. In particular, the physicists who had been working on the atom bomb during the war had returned to the universities and were very eager for basic research.

Of course, I had all sorts of problems getting there; for example, I had to wait nine months for my visa because the CIA had tainted my father as a communist.

That fall my wife and I boarded ship in Holland and had just six days to become adjusted to the different time zone and the different country. The trip was not a dream cruise as the Atlantic was too stormy at the end of October for that, but we arrived in New York on time. To be precise, we arrived somewhere south of Hoboken and that seemed to be the end of the road. The situation was completely new to us: there was a strike going on that had brought all public transportation to a halt. We stood there with our 16 suitcases on the pier and didn't know how we would be able to continue our journey. While I was looking around trying to find out what the other people were going to do, I saw two young guys hanging around a car. They made a relatively trustworthy impression, and I asked them if they wanted to earn a little money. I didn't have to ask twice. They immediately said yes and loaded our suitcases with us in their car and drove us to Princeton. They let us out at the train station.

We had arrived at a place where I didn't know anyone, or my way around. The only thing I had was the telephone number from Oppenheimer's secretary, but I didn't really dare to hope that I would be able to immediately broach this sanctum. After all, it had taken more than a day to arrange to speak to Heisenberg in Göttingen.

Thankfully, here everything worked out fine. The secretary immediately came to pick us up in her car, gathered all of our things and drove us to the house that was provided to us to stay in. We were overcome with joy when we saw that the refrigerator was filled with all the necessary items! Oppenheimer had really organized the ideal ivory tower for his guests.

It was just as easy to become accustomed to life and work here. I immediately got to know colleagues with similar interests, and because we spoke the same language and had the same background, we could begin work right away. Before I get to that, though, I'd like to talk about my meeting with Einstein.

First I'd like to describe the Institute, which was explicitly conceived for Einstein. A businessman from New York felt the need to give something back to the country which had made him rich. So he wanted to found a scientific institution which could offer researchers ideal working conditions. His guests were carefully chosen but did not have any obligations and could devote themselves completely to their work. He did not spare any cost to make the Institute sufficiently attractive to hire the biggest draw. Naturally, Einstein was his first choice.

His plan worked. Einstein came and the Institute was the Mecca of physics for decades. Einstein was, of course, retired during my time there (1953–1954), and he died in 1955, but he continued to come to tea from time to time and find out what everyone was up to. He had an impressive appearance. I can still see him before me as if it were yesterday: the white mane of hair was somewhat thinner, his face furrowed by the many successes and failures, but his charm remained intact.

As a child, he was late in learning how to speak, and he also had to change schools when his father had to move after his electronics business began having difficulties. The business eventually went bankrupt and Einstein had to struggle through all sorts of tutoring and odd teaching jobs to finance his education. Even after graduation, he wasn't able to find a job in academia. Finally, through personal contacts, he found a job as a third class civil servant in the patent office in Berne. This is where his peerless success began.

He found friends who shared his interests. In their discussions, he was able to discern things that remained obscure and hidden to the greatest intellects of the time. Most of the great innovations in physics in the 20th century are based on Einstein's ideas. On the other hand, his commitment for peace was a complete failure, although his writings about peace fill more pages than his writings about physics. He fought for peace in countless committees and organizations, but it came to naught. No one could stop Hitler's path to war.

I was fortunate enough to get to know Einstein personally. He felt a bond to my father, who had been one of the few people participating in the international protest when Einstein was thrown out of the Prussian Academy of Sciences. Our first conversation took an unexpected turn. After I explained that I was just an associate faculty member in Berne, his face lit up and he raved about what a wonderful time he'd had in Berne and how much he'd learned about physics during his time there. I could understand the first part, as Berne was where he'd enjoyed the first years of his marriage to Mileva, but the last bit surprised me. Of course I knew that he'd published three revolutionary articles in Berne, one of which later won him the Nobel Prize and the other two should have done the same. However, at the time there wasn't anyone in Berne of his caliber whom he could have learned anything from. Einstein put it like this: he used to like to walk through the old town to the bear caves and watch feeding time. He observed that the bears usually walked with their mouths to the ground and would only find what was in front of their noses. Sometimes one would get up on its hind legs and could see from this higher perch where the really good treats were. This reminded him of most physicists who bent over their calculations and only saw the last equation. But the most important connections are discovered only when you can see the situation from above.

While we were in Princeton, my wife and I were twice invited to Einstein's house at 112 Mercer Street for sandwiches. We mostly spoke about politics. We were witnessing the apex of the McCarthy era and Einstein was deeply troubled by the witch-hunt climate. For me it was more grotesque. I once heard the following at a "hearing": a black cleaning lady in a government laboratory was being questioned for having communist ties, and asked if she knew a Mr. Smith. She replied that she did. In fact, there was a notorious communist named Smith and it seemed that she had been caught guilty as charged. However, it surfaced that the communist Smith was white, but the man she knew was black. One would think that the charges would clearly be dropped, but not with McCarthy. He simply said that that was a minor detail to be dealt with later. In any case, the communist ties had been proven. These practices were painful for Einstein, having hoped as he did after the Nazi terror to at least find consistently more humane and dignified conditions in the USA.

What I noticed as Einstein's most marked characteristic was his simplicity. This already began with his clothing; usually he wore just a sweater and a wrinkled pair of pants. Socks were for special occasions.

In a similar vein, his speech was also simple; he never attempted to impress others with ornate language. His entire appearance was not that of a high ranking official, but maybe that has only served to strengthen the lasting effect he has had. He owed a great deal to his genius of a secretary, Helene Dukas. She controlled his public relations and blocked a lot out, but not everything. Some things she allowed to slip through. So once a little boy wrote and asked Einstein for help with his math homework in a letter, and Einstein did it. Helene Dukas made sure that this remained the exception.

My wife and I became friends with Helene Dukas and continued to see her often after Einstein's death. In this way we learned a lot about her famous boss firsthand. Once, for example, my wife complained that our three-year-old son was still having problems talking. She simply said, "Don't worry, the Professor (she was, of course, referring to Einstein) only learned to talk when he was four".

Naturally, Einstein and I often talked about physics, but he didn't want to believe in the quantum field theory that my generation was working on. I once told him about particle creation through a gravitational field, an effect that had to exist and would later become popular as the Hawking radiation. At first Einstein didn't want to understand what I was talking about, for he thought I was talking about cooking up new nuclei inside stars. When I made it clear that this was a pure transformation from energy into matter he was shocked and said that he would never accept such a thing.

He was looking for a purely classical field theory which would unite all forces. Today we know that it was the wrong direction, nature works differently, namely quantum mechanically. Only much later did I discover that Einstein had been working on another phenomenon while at Princeton, one which is now just becoming meaningful in astronomy, the gravitational lens effect. According to Einstein's theory, a gravitational field bends a ray of light so that it seems to be attracted by the heavy body. A star then functions like a converging lens, albeit one with poor optic quality. Still, it makes it possible to see massive objects with a weak luminosity, particularly black holes. The zeitgeist wasn't ready to think about such things and Einstein remained alone in his science. Neither he nor anyone from his circle spoke about this work at the time.

Besides Einstein, there were other researchers at the Institute who made pioneering breakthroughs, and I'd like to mention a few. They show what kind of intellectual capacity the Danube monarchy was capable of producing.

One of these was Kurt Gödel. He was from Brno and studied in Vienna. This was in the 1920's when mathematics in Germany was dominated by David Hilbert. Hilbert wanted to close the book on mathematics using an axiomatic method. The way it would work was that first the meaning of the terms would be determined through certain axioms, and everything that is true should be able to be derived through logical deductions from the axioms and therefore provable. Then you'd have everything which was true and false in black and white.

While as a young freelancing assistant at the Wiener Institute, Kurt Gödel shattered Hilbert's triumph. He showed that such a program could not work. There are always going to be statements that are not provable, but the opposite is also not provable. Hilbert's idea of black and white didn't hold; there are too many statements that are neither true nor false, but undecided.

Fig. 14: Kurt Gödel

Gödel's oft-quoted theorem has an enormous effect that goes beyond pure mathematics. It changed the way cognitive science worked. You'd think that after having proven that the emperor was wearing no clothes, Gödel would have been offered a professorship. This was not the case and he continued to eke out his existence with the fees paid to him by two or three students. His value was appreciated and recognized in Princeton, and he was there when Hitler came to Austria, where afterwards there was the big "cleansing" of the university and Gödel was also thrown out. Still, he returned to Austria and was astounded to find that he had to be dismissed for being Jewish. He made the counterargument that he wasn't Jewish. This embarrassed those in charge and they said that in that case, he could continue giving lectures. He had become more self-confident in the meantime and said that he deserved a professorship. The reaction to this was that if he wanted one, he should turn to the employment office. It dawned on him that he was in the wrong place, and that these were the wrong people for him, and he returned to Princeton. This was not so easy, as the war had begun and the direct route was cut off. He had to take the more difficult way via Siberia, Vladivostok, Japan, Hawaii and California. He had Adele Porkert to thank for the fact that he made it. She was a dancer eight years older than he, whom he had married against the objections of his family.

In Princeton he lived very reservedly and socialized only with Einstein. In the Institute they said that Gödel had to arrange to get American citizenship, and as he was very impractical in everyday matters, Einstein was helping him with it. One evening, Einstein came to Gödel and brought him a copy of the American constitution, saying that he just had to read it through and the next morning they would drive to Trenton where he would have to take an oath on it, fill out and sign a few forms, and that would be that.

The next morning Einstein found Gödel very upset and he said, "Look, Albert, I just can't do this". Einstein, taken aback, "Okay, why not?" Gödel said, "Like you said, I read everything through and noticed that the whole thing is not logically consistent. How can I take an oath to something that's not logically consistent?" Escorting someone like this through all of Siberia during a war winter is truly an achievement, and I have profound respect for Adele.

In Princeton, Gödel's genius once again could flourish as he turned to a completely different area, Einstein's gravitational theory. It was always said that the theory of relativity had done away with absolute time. First, and this should be emphasized, it's important to see that it's the other way around. There aren't too few variables that could serve as time, but too many and none of them is preferred by nature.

Gödel discovered a model of the universe that in fact did that which had only been a rumor before; in the entire universe no time is definable, because you can fly back into your past. There you could cause all kinds of trouble, like even destroying your own basis for existence — therefore another logical inconsistency!

Personally, I only saw Gödel from afar at teatime. Somehow I hadn't caught on to his importance and didn't really try to approach him, so that I don't have any personal memories about him to share.

The other scientist I would like to mention here is Johann von Neumann, an unbelievably brilliant man. He was officially a mathematician, but in fact, he created many new branches of science. He forged the mathematical equipment for quantum theory, he initiated modern entropy theory, and he is the father of game theory. He was also responsible for practical but unfortunate things, such as how to ignite an atom bomb: two uranium balls under critical mass which together are more than the critical mass need to be shot at each other.

During the time that I spent at the Institute, von Neumann had just isolated himself because he had recognized that the future would belong to the electronic computer. He insisted that one be built at the Institute. The other mathematicians felt that this was undignified waste of time, and von Neumann and his computer were banned to a kind of barn as far away from the Institute as possible. At the time, the transistor had not yet been invented, and electron tubes were used in electronics. These generated heat like a light bulb, and as you needed thousands of them, the biggest problem was how to get rid of this heat. The largest amount of time and money was spent on air conditioners. What a chip the size of a thumbnail can do today would have cost an entire city full of air conditioners and this was nothing for the dreamy ivory tower of the Institute for Advanced Studies.

As opposed to Gödel, "Jonny" was actively involved in Princeton's social life. My wife and I had the pleasure of being invited to his parties. He turned out to be a charming host and always enjoyed dancing with my wife.

Besides these giants, there were also three permanent physics staff members who all left their mark on physics: Freeman J. Dyson from England, Abraham Pais from Holland and Chen Ning Yang from China. The stream of visiting researchers who came for limited periods was just as international, and also made important contributions to 20th century physics. There was; for example, Leon van Hove from Belgium. He would later play a leading role in CERN. It is still an open debate whether he or Victor Weißkopf was the best director that CERN ever had. In any case his wife Jenny was the undisputed best "first lady" of CERN.

This is not insignificant. Personal relationships are important for the climate in a group that's so internationally mixed.

The Swiss protagonist in Princeton was Res Jost, he would one day become an internationally renowned figure on integration in the area of mathematical physics. His wife Hilde was also a pillar of society. She was Viennese and had been Felix Ehrenhaft's secretary.

Scandinavia was represented by Gunnar Källen from Sweden, who was a completely different type of person. He defended his position as Pauli's favorite to the bitter end and scorned everything that would have threatened his rank. However, this aggression was limited to science. When he stayed in Princeton for a longer period and we invited him to stay with us, he was a very pleasant and helpful guest. His scientific brilliance was beyond doubt, he never made a mistake in his calculations. Unfortunately, one time he made a mistake in another area, as an airplane pilot, and they can only make the one.

Japan was also represented by Toichiro Kinoshita and Yoichiro Nambu. As opposed to the Europeans mentioned, they remained in the USA as the rebuilding of Japan was taking too long for their taste. Or was it taking too long until Japan recognized the significance of these scientists? In 2008, Nambu received the Nobel Prize in physics.

The USA was represented by Stanley Deser, Boris Jacobsohn and Fritz Coestler.

The combination of nationalities at the Institute was balanced; in general there was a wide array of possible sounding boards. Perhaps because there were so many others I missed my chance to come into scientific contact with giants like Gödel or von Neumann, because for us younger researchers, there was only one area we wanted to do research in, and that was quantum field theory.

Quantum Field Theory

Around 1930, soon after the laws of quantum theory had been established, a group of young theoretical physicists (Pauli, Heisenberg, Dirac, Klein, Jordan, and Wigner) were able to unite them with the principles of the special theory of relativity into quantum field theory. As I mentioned before, this brought about a complete revision of our understanding of the nature of matter. Up till then, the elementary parts of matter; protons and electrons, had been thought to be tiny balls whirling about. The space between them was understood as being empty, so actually in fact nothing. Mechanical models were sought to explain disturbing factors like the electric field, and they were the only acceptable paradigm.

Now there was a radical change in perspective, and the field became the fundamental term. The field fills all space and is the carrier of the laws of nature and of matter. Particles in the classic sense are just local stimuli in the field; they can come into being and cease to exist; all particles belonging to the same field are indistinguishable, as if they had been cloned. If the equation sufficient for the fields is invariant under the Lorenz transformations, then the alliance between the two schools of thought would almost be complete. There was just one fly in the ointment. The theory of relativity strictly forbids anything spreading faster than the speed of light, but quantum field theory needs ever higher levels of energies and impulses to observe such sharp boundaries. The reason for this lies in the wave nature of matter. The sharper something is localized, the shorter the wavelengths needed to do so. In quantum theory, short wavelengths pull high impulses along behind them, and high impulses are what are presumed in this kind of theory. This goes so far that when you want to calculate something in quantum field theory you will soon receive "infinite" as an answer.

However, clever people then discovered that the infinities can be seen as an addition to a particle's mass and charge. There is a way to get out of this plight. As the original values of mass and charge are unknown, the values measured can be taken for their original plus the above-mentioned additions, and their scandalous past is forgotten. This is called renormalization.

These operations can only be carried out in steps, the addition being represented as the sum of the effects with a certain number of interactions and each summand must be treated individually.

When I came to Princeton in 1953, I had already proven that such a sum can never converge as mentioned above. Therefore, it was a point of honor for me to avoid this kind of series expansion. In Princeton, I worked with Marvin (Murph) Goldberger, a brilliant physicist and mathematician, and we racked our brains trying to find a way to get results from field theory without using an infinite sum. Murph told me that he and Murray Gell-Mann had once derived relations between the scattering probabilities of electron–photon scattering under the assumption that nothing could spread faster than the speed of light. It was only that neither of them knew how they could derive this from quantum field theory. In the standard formalism, the so-called Feynmann propagator was used in propagating effects, and this was not exactly zero for points outside the "light cone", which could therefore only be connected faster than the speed of light.

Fig. 15: Murray Gell-Mann (on the right) and Walter Thirring

Something there rang a bell, as I realized that events that could only be connected faster than the speed of light had to be independent of each other. In quantum theory, when a is independent of b then ab = ba. So it could be correct to say ab – ba = 0 when a and b lie outside the light cone.

Murph and I clung to this notion, and we were, in fact, able to cultivate the desired relations out of this meager information. However, at first we had taken the easy route and only considered those particles which were also mass points. At that time, the only mass point particles were the π-mesons, but these didn't live long enough to be scattered around on each other. To be able to confront our predictions with experimental data, we had to return to electrons and photons. These are like little tops, the electron has a spin ½, the photon spin 1, and this obfuscates the calculations.

Murph suggested the following: Murray should work this out and we would include him in the publications. Murray was a young genius, two years younger than I, and was considered to be a man of great ideas. At first, I was a little skeptical if we could simply throw him the dirty work. Imagine my surprise when two weeks later we received a wonderful manuscript from Chicago, where Murray was working at the time. Murray had not only immediately comprehended our demonstrations; he had also transformed the supposed dirty work into an attractive mathematical theory. This once again proved to me that the important thing was not all the debating in cafes, but the perfect mastery of technique.

Berne 1954–1956

Although Oppenheimer asked me if I wouldn't like to stay in Princeton for another year, I declined, as I thought to myself that you don't simply give up a permanent post in Switzerland. In Berne, my position was low in the academic hierarchy, but there were other advantages. I've mentioned before that the mood barometer of my boss Houtermans was not always pointing to "fair", but I'd already gotten used to it, like you do to the weather. The important thing was that he took care of his people, which he did. So that I would not be intellectually isolated, he paid for me to go to Zurich once a week so that I could attend the Pauli seminar. He also radiated humor and wit in the institute where it was readily apparent that he was not exactly reserved in his choice of words. For example, once he couldn't remember the name of our meteorologist, and he simply said "you know, the storm hag".

I was completely free to do science, and the genius loci suggested that I again return to the theory of relativity. My first work had been dedicated to it, after all, and even if it wasn't a big breakthrough, it still showed my contemplative style in physics. In any case, I thought that I understood the special theory of relativity; in the general theory I could repeat the usual, but a lot of things were meaningless to me. For example, I couldn't really do anything with the principle of general covariance, because — Einstein or not — why should someone prevent me from introducing my favorite coordinates? I also noticed that sometimes the tone becomes haughty in the general theory of relativity. "Gravity doesn't exist, only space and time are curved and this gives the illusion of gravity". To do away with all of Newtonian gravitational astronomy, the crowning glory of physics for centuries, as just an illusion was not something I enjoyed, particularly as this statement would call into question the reliability of every scientific result. I wanted to set things right a bit and understood that the other extreme position, namely that gravity changes length and time in a way to create the illusion of curved space, is also logically defensible. Both provocative statements suffer from the ontological question of what is illusion and what is reality. Due to a lack of further meaning, it's better to leave the question unanswered and to choose the wording "gravity curves space and time" as a compromise.

Curved Space

I was therefore trying to generate the geometric meaning of gravity rather as a result; not plugging it in as a presumption at the beginning, but coming up with it at the end. This idea was met with both approval and disdain. From the latter I will only mention Einstein, with whom I'd earlier discussed this idea. He thought that when you only argued using field theory, you first had to consider the simplest case; anything else would be a "sin against the Holy Spirit". The simplest thing would be a single gravitational field, a scalar. Today we know that there are 10 interwoven gravitational fields which together form a tensor and this conclusion could only be reached with the geometric point of view (Einstein's view).

At that time, I didn't have a sufficient overview of differential geometry to be able to respond that the simplest Riemannian geometry is conformly flat and also only has one field. But also the Riemannian geometry that Einstein used is not the most general, the so-called torsion is set at exactly zero in it. From a purely logical point of view, Einstein's approach is not the only possibility, but as we know today, it's exactly right. Einstein seemed to have somehow had a direct line connecting him to the man above!

Still, it seemed to me to be desirable to first deal with gravity like every other force, and leave it alone to emerge into its special role. The point was to build up the general theory of relativity (= theory of gravitation) like the special theory of relativity (my view):

Step I

In discussing the consequences of a theory of gravitation which obeys the axioms of the special theory of relativity, it becomes apparent that:
a) bodies expand due to the gravitational potential
b) in doing so their movements slow down. This happens at the same place for all bodies and by the same factor, which can have different values for different points in space time.
c) the speed of light is the general speed limit, but has different values at different points in space time.

Step II

If the observed lengths and times are introduced to the metric, then we can see that the equations of motion in such a theory are identical to the equations without gravitation in a curved space.

Step III

As the original Minkowski space remains unobservable, it is forgotten, and we can be daring and coolly say:

The space is curved, but there is no gravitational field.

Sometimes in common usage we forget to clearly differentiate the unobservable Minkowski space and the Riemannian space we discovered. For example, Step III says about the redshift in the spectral lines in the light of a star that identical atoms all have the same frequency at any spot — they define time. On its way to us, the light needs to go through different metric zones, which changes its frequency. As seen from Step I, it's clear that the gravitational potential (not only the field strength) influences the size and frequency of the atoms, but in a static Minkowski space, the frequency of a ray of light never changes, and there is no other redshift. The consequences observed, namely the change in frequency between the exiting ray and the local atoms, are the same in both cases.

Perhaps for everyday use it's enough to use the more casual formula, just like we also say "the sun is rising" and not "the Earth has spun a bit more so that we can now see the sun". It's just important to remember that these are two different descriptions of the same effect. For example, you can't say that the atoms move a little slower on a star and that the photons get a little tired when they move out of a gravitational hole, both effects can be explained, therefore they should be added together. This would produce one factor of two too many, and you can no longer afford that when you're dealing with a standard of precision reaching into several thousandths. This plays an important role particularly for a GPS and Step I is used in the brochure where they say that due to gravity the clocks on the Earth's surface move more slowly than those up on the satellites. Of course, GPS users just want to know where they are and aren't interested in ontological discussions. Nonetheless, as a physicist, it's important to be precise in what you're talking about and which point of view you've embraced.

However, at that time I didn't have any students in Berne who could have adopted my ideas. This was to change in Vienna, where the intellectual climate had been favorable towards the general theory of relativity from the very beginning. Right after 1916, when Einstein published his general theory of relativity, there was a group of theoreticians in Vienna; including Schrödinger, Pauli, Flamm, Kottler, and Hans Thirring who eagerly embraced the theory and improved on it with their own contributions. The war robbed them of their leader Friedrich Hasenöhrl when he was killed in southern Tyrol, and they had to work under the most difficult circumstances. Still, they soon were able to fight to the very top to become world class in this area. Also, after I had returned to Vienna and revived the tradition of giving a lecture on Einstein's theory, Roman Sexl, my first student, was immediately inspired and began to build up a group about the general theory of relativity. Many talents resulted from this group including Peter Aichelburg, Helmut Urbantke, Helmut Rumpf, Bobby Beig, and Franz Embacher. All of them delivered fantastic ideas about gravitational theory. In the meantime, these ideas have even spread out and become popular among teenagers outside of the big cities. For instance, Cornelia Faustmann was fascinated by the theory of relativity as a 10 year old in the small town of Wiener Neustadt, and began to study it with the result that she was able to write an overview about it when she was 17. But all of this for me was still in the distant future.

MIT and the University of Washington (1956–1958)

My work had generated so much resonance that I received invitations from many different directions. One was too tempting to resist, namely to the Massachusetts Institute of Technology and to Viktor Weißkopf. In democratic America, the standing and reputation of all the different universities is by no means the same. This is still true today, although now that there is such a large number of excellent researchers, you can also find leading scientists at smaller universities. There are, for example, many universities in and around Boston, but Harvard and MIT have the most appeal. The best students from all over the country come here, which in turn attracts the best professors, reinforcing the standard of excellence. I also experienced how the financially strongest families enjoy sending their offspring to these prestigious universities when, once in Princeton, as there was no house available for me, I was given quarters in a dorm for graduate students. At first I was a bit wary, but then after receiving such a comfortable two-room suite I even complimented the university officials on their treatment of their students. To my surprise, I was told that this housing wasn't even intended for students, but for their servants instead.

At first, MIT seemed to me to be like a beehive, the buzz of hard work could be heard everywhere. Therefore, I wasn't so sure that I would be able to find the necessary peace and quiet to be able to think straight. This turned out not to be a problem, as it was exactly here that I began the work that would make my name in the physics literature. Due to the general noise level, I enjoyed going to the library. Here, you weren't allowed to speak, and you could look through the journals without being disturbed. It was here that one day I came across an old article from Hans Bethe in which he solved the following dynamic problem: he observed an infinitely long chain of spins, small tops in which each one is welded together to its neighbors on either side. If you turn a spin out, the perturbation propagates itself like a wave through the entire chain and the most general motion is an overlap of such "spin waves".

I realized this was mathematically similar to what we were doing in elementary particle physics when we observed Fermi particles. These are the particles with a half-integer spin. Heisenberg had just coined the phrase "everything spins half", by which he meant that he could reduce everything to spin ½ particles, with interaction only taking place between direct neighbors. Naturally, he was not only thinking of a chain, but worked in three dimensions. This is why he could not solve his equation, but just offer something that he considered to be a good approximation. Therefore, I thought it was important to offer an exact solution in a simplified case in order to check how good his approximation was. I set about my work and worked out Heisenberg's equation in only one spatial dimension with the Bethe ansatz.

In fact, I was able to reduce the problem to an equation with just one variable, which was already progress considering I'd started out with an infinite number of variables. However, this equation was an integral equation, which doesn't mean an equation containing an integral, but that the value you're looking for appears under an integral sign and should be integrated.

I thought to myself that I'd be able to work this out as well, especially because I'd been invited to Princeton for a few weeks, where the best mathematicians in the world were. So that I wouldn't have to toil unnecessarily on something which was probably already known, I went to a famous mathematician and showed him the equation. He listened to me patiently, made a few small changes in the notation and praised the equation. Everything was well defined and I should be able to solve it easily. However, exactly how I should be able to he couldn't say.

The next mathematician was even able to prove that the equation had a solution, but his proof wasn't constructive, it didn't say how the solution could be found. I decided to stop soliciting the mathematicians and threw myself with everything I had into working on the equation. I took it apart and introduced other variables until suddenly the solution stood before me. Was I proud of myself! Years later, I found out that the equation had been solved long before I had done so, and in some circles was considered to be a particularly simple type.

In any case, I set about publishing this result. It was just at the right time, as a few years earlier T.D. Lee had published a solvable model of quantum field theory, but this did not fulfill the requirements of the theory of relativity. His model had already been exploited, and lots of unemployed theoreticians were waiting to finally receive a Lorentz invariant model. My model provided exactly that feature.

Still, some experts were skeptical. Harry Lehman said that a Lorentz invariant quantum field theory is something so complicated that an exact solution is unthinkable, something couldn't be right. Källen groused, "just more sloppy work from Thirring, not all of the questions have been answered". Since today, a half century later, new features in the equation are still being discovered, I accept the last criticism as being correct, but not so important.

Luckily, Jurko Glaser immediately recognized that the equation could be shown in a closed form. For lack of a better word, he called the whole thing the "Thirring model" and this name stuck. I owe the fact that my name continues to be alive in the literature today to this circumstance. The model currently has a Google number of 178,000; that is the number of entries Google finds in 1/3 second. This is a good example of how a random coincidence can build a bridge between failure and success, between obscurity and fame.

In 1956, there was a surprise development in physics: parity violation had been discovered. I will describe the situation briefly, because the reactions of famous intellectuals to something they hadn't expected are always amusing.

The thing about parity violation is whether nature prefers one helicity to the other. In physics, the reverse of a space direction is the operation P, "parity". Naturally, it had always been assumed that right and left were equal, without questioning this assumption.

Excursus on Parity

P can be realized so that the x-coordinates do not run from left to right, but from right to left. Using P twice leaves everything unchanged, so P.P = 1. Simple quantities, "scalars", which are only multiplied by one number when P is applied, are subdivided into scalars and pseudoscalars. This number squared must result in 1; it can only be 1 or –1. The word pseudo is meant to signify this change in sign. That such pseudoscalars exist in nature was discovered with great cunning. The so-called π-meson was one of these peculiarities.

Now there were many heavy particles that decayed into the lighter π-mesons. In particular, there were some that decayed into two and some that decayed into three π-mesons, but when the mass was calculated out of the decay, they seemed to have the same mass. Did this mean that two different particles just happened to have the same mass? This would be very odd, because otherwise different particles have different masses. Were these perhaps one and the same particle? But what is with its parity? Two π-mesons have parity $(-1).(-1) = 1$ and three have $(-1).(-1).(-1) = -1$. Something must have gone wrong here.

At first, the particle was dismissed as a schizon, but then someone came up with the idea that parity might not be conserved during decay. I had the opportunity to talk to the most important scientists involved before the answer was known and received the following reactions:

Richard Feynman: *"Maybe it's the case that the parity is almost conserved and that's why the decays take so long".*

Abraham Pais: *"I accept it only with great reluctance, but if it has to be, then it has to be".*

Markus Fierz: *"That would be extremely odd, because then space would be screwy".*

Frank Yang: *"I think that it's impossible".*

Everyone thought that this kind of screwiness in nature would have to have been noticed by now, but it was only Yang together with T.D. Lee who got to the bottom of it. They discovered that what had been measured before were scalars, quantities that do not change with P. They do not provide any information as to whether a violation of parity was signalized. For that, a pseudoscalar would have to be measured, and they urged for these measurements to be taken. A half year later, these were measured by the Chinese physicist Wu with collaborators and parity violation had been discovered. One year later Wu, Yang and Lee received the Nobel Prize for this discovery.

After I had spent a year on the east coast of the USA, I thought I'd like to get to know the west coast as well. The invitation from my friend Boris Jacobsohn to be a visiting professor at the University of Washington in Seattle was therefore very opportune. Before this could happen, I had a few logistical problems to solve. Our older son was already two and my wife was pregnant again and not so mobile. I decided that I would put her and my son on a plane in New York and drive after them in the car. My wife agreed without a grumble, just as she was always willing to follow me along the labyrinthine path of my career. For me, the drive from coast to coast was an unforgettable experience. It wasn't as big of a challenge as you might imagine; the past year I had had to drive through Boston twice a day in rush hour and had therefore gained quite a bit of driving experience. Also, traffic in the USA is much more disciplined than in Europe, and therefore driving is less stressful. I was able to truly enjoy the trip; the beautiful and varied landscapes formed a magnificent background. There was something new every day; in Yellowstone Park bears and bison came up to the car, in Glacier National Park you could marvel at the climbing abilities of the mountain goats. At the end I crossed the Cascade mountain range, and finally reached the ocean in the west.

The university in Seattle was in the process of becoming established, and it was hoped that the many opportunities the city had to offer would attract the top people. In the meantime, this hope has been fulfilled and the university has developed into an excellent institution. I was soon faced with the decision about what I wanted to do with my life. I was offered a full professorship here. On the other hand, I had heard rumors that my name was also on some lists in Europe, including that of the University of Hamburg. However, there I was the second choice after Harry Lehmann. He accepted the position, so that the decision had been taken out of my hands. Still, soon thereafter offers came for professorships from Frankfurt and Munich. I had to figure out what I ultimately wanted.

At that time, my father retired and his position had to be filled. It was in some way tempting to succeed him, although this was by no means the most attractive possibility; it was even perhaps a risk. I didn't have an offer to go to Vienna yet, just my associate professorship in Berne. The other positions were also awaiting my answer. Therefore, I had to turn down a full professorship in Frankfurt from my position as associate professor in Berne. This went against the unspoken rule that you could not turn down your first offer as a full professor. You won't be locked up if you do, but scorned. My main mentors, Schrödinger and Pauli also advised against it. Schrödinger said that experimental physics had never really been able to establish itself in Austria, and I would not have any contact to the experiments there. Pauli thought that there was a general unscientific atmosphere pervading Austria, and I would not be able to make any further contributions to physics there.

It turns out that both were mistaken. Schrödinger didn't think about the fact that times can change, and Pauli had extended his troubled relationship to his father to include all of Austria. In my visits to Austria I had noticed that there were very many talented young people. They just needed self-confidence and guidance toward the right problems. It's hard to pinpoint what exactly swayed me at the end to go for Vienna. Did I have a feeling that there would be a Rauch, a Wess, or a Zeilinger among the talented young people, or did the disparaging opinions of Vienna arouse my spirit to prove them wrong? In the final analysis, it might have been my sense of family duty which made my decision. After all, I had seen that my father had made no further significant contributions to theoretical physics after 1920, and that, therefore, Vienna was considered to be a wasteland in this area. I can remember a letter from my grandfather to my father in which my grandfather charged him with making sure that the name Thirring have a prestigious ring to it throughout the world, and this is exactly what I wanted to do.

Therefore, I turned down offers from the USA and Germany and returned to my standby position in Berne.

Berne (1958–1959)

On my return to Europe, I could feel a change in the mentality in scientific circles. Up till then, the shock of the Second World War had not yet been overcome and the dominance of the superpowers USA and USSR in physics was only too obvious. This was particularly true for elementary particle physics, because the large particle accelerators which allowed for the production and examination of the massive, yet instable particles were unfortunately not located in Europe.

The best European physicists worked in the USA and it was clear to them that the intellectual potential for research was available in Europe, but the necessary resources were beyond the means of the individual countries to finance. But now the politicians had realized that when all of the countries pooled their financial resources together, it would be possible to become competitive by operating a *European* laboratory. In 1954 CERN was established, and Geneva was chosen as the site. In fact, I was present at this historical event. We were driven by bus from Geneva into the middle of nowhere; there was just a huge field. It seemed presumptuous to think that here we would compete with intellectual centers like Berkeley, yet there was a spirit of "together we will be strong" like the oath of Ruetli among the representatives of the different countries. It was a first sign of European unification.

While it's true that Geneva is expensive, at the same time it offers many other advantages. Houtermans immediately understood the significance of CERN, and knew how to capitalize on the proximity between Berne and CERN. I received permission to have off two days a week from my associate professor position in Berne so that I could go to Geneva and work there. In the group for theoretical physics in CERN, I not only met the leading experts from Europe, but CERN had also begun to attract scientists from the US, Japan and many other countries. Of course this meant a six hours' drive for me each week, there weren't any highways at that time. The undertaking not only allowed me to profit scientifically; I was able to fully enjoy the beautiful surroundings of CERN. To the west begin the French Jura and there was a small cable car to the mountains located in the nondescript town of Crozet. Close to the mountain station there was a tiny guesthouse where you could stay the night. I arranged for my accommodations there and could enjoy a magnificent panorama every day. Mt. Blanc Massif was located vis-a-vis; a spectacular view. In the evening, the foreground sank in darkness but the peak continued to shine for a long time until its light slowly became extinguished. In CERN, people thought I was a little crazy for going up to my mountain every evening, but in the morning when I could enjoy looking at Mt. Blanc at breakfast and saw how the people below were covered by fog, I thought that actually they were the ones who were crazy.

Once the large accelerators started operating, the discovery of new elementary particles began. Other than mesons, which are the particles with a mass between that of an electron and a proton, particles were found that were heavier than a proton. They were called baryons, but more important than finding them a name was finding a way in which they could fit smoothly in with everything that was known up till then.

A new "space" was opened, the inner space which is hidden from us when viewed from outside. In this space, every kind of particle carries one dimension and these directions are connected by symmetry relationships. To put it mathematically, transformations of the types of particles build a group amongst themselves. Group means that two transformations occurring one after the other also result in a symmetry operation. At first, the group needs to be found which describes the symmetry of the situation. For the baryons already known at that time, i.e. protons and neutrons, it was also recognized that this was group O(3), the group of those transformations of a three dimensional space in which orthogonality is conserved. The group that contains other kinds of particles must also include this group; to put it mathematically, to include O(3) as a subgroup.

This is why Abraham Pais first tried his luck using O(4), four dimensional rotation, but Providence was not smiling down on him. I took a different path and assumed that O(3) could also be interpreted as U(2), the group of rotations in the two dimensional complex spaces (unitary transformations). Schwinger had always emphasized that the natural symmetry for particles with a half integer spin were unitary transformations, and for particles with whole integer spin, it was orthogonal transformations.

The newly discovered particles proved the existence of another quantity found, that of strangeness, which suggested expanding U(2) to U(3). This means that all baryons were built out of three basic baryons and the mesons were baryon-antibaryon pairs of which there were $3 \times 3 = 9$.

In the meantime, it's been shown that the meson classification was correct, but our assumption that the basic baryons were the proton, neutron and the newly discovered lambda was wrong. Only several years later did Murray Gell-Mann and George Zweig discover that the so-called quarks carried out this task. You cannot observe them individually, they are bound together in groups of three, and this conglomerate is what we see as proton, neutron and lambda. When I say we, I mean that I was working with Julius Wess, a young Austrian, at CERN. He had a scholarship for a period at CERN and was very gifted in mathematics. He was largely an autodidact and came from the rural Sölktal valley in Carinthia. He would become one of the leading experts in theoretical physics of the second half of the 20th century. He was taken by my view that the dynamic was largely determined by symmetries, and soon he had come up with all representations of the U(3) group.

However, our studies did not fit in with the spirit of the times. No one else at CERN was interested in symmetries of the Lagrange function because the Lagrange function itself was seen as outdated. We only had one contact in Japan, where people were working in this direction. Therefore, my lecture on this at the conference in Kiev in 1959 was met with little understanding. For instance, Landau asked me how come I knew that this baryon-antibaryon structure was the π-meson. I answered, "because it has the exactly right quantum number" but could not convince him. Heisenberg was already made uneasy by expanding U(2) to U(3). This seemed to him like a messenger to a strange world that didn't fit in with his views. For reasons I'll shortly discuss, I didn't have a lot of time to argue my point, so that this work lay on ice until Gell-Mann got on board two years later. In the meantime, more substantial empirical material on baryons had been collected, and the new picture was a perfect match. Our mistake was that we didn't have the guts to see all of the eight lightest baryons as being connected by symmetry. Julius learned his lesson and together with Bruno Zumino he connected bosons and fermions, which are the extreme cases in the opposite direction, and they became the world famous founders of supersymmetry. This became the basis for the areas being most actively worked on today, like string theory or M-theory.

Soon after I had returned to Europe in the spring of 1958, I was offered a full professorship in physics and the directorship of the Institute for Theoretical Physics at the University of Vienna. In the fall of 1959, after 1 ½ years of negotiations, I began my work in Vienna. During this time, I would always have to drive to Vienna for meetings, which was difficult and extraordinarily depressing for someone who had spent a long period of time in the US and in Switzerland. The autobahn existed only in fragments, so that I needed two days to drive from Berne to Vienna. Austria at that time can hardly be compared to Austria today; after all, it was only three years old. The signing of the Austrian State Treaty took place in 1955 and the development from a rubble pile to a wealthy industrial nation had not yet been effected. There was still much in chaos, and due to the fact that the necessary financial and legal conditions were still missing, these deficiencies could only gradually be remedied. I would like to take the opportunity to briefly set out the basic problems and their solutions, as it might be interesting for the modern reader to see which paths lead to success or to failure.

The problems:

(1) The building was filthy and rotten, and certain minimum standards such as toilet paper were not being provided.
(2) The Institute was not being equipped with sufficient personnel, there were only 2 ½ assistants and there was no secretary.
(3) The university curriculum was missing many modern branches of study such as quantum mechanics.
(4) Austria lay isolated at the edge of the Iron Curtain, and was therefore not incorporated into the international scientific infrastructure.
(5) Austria was about to enter CERN, but experimental high energy physics wasn't being done anywhere in the country, so that the large expenses didn't make any sense.

My conditio sine qua non was that these serious deficits be resolved, but I did not suffer the illusion that the Ministry for Science would be able to do so. I would have to come up with the money from somewhere else and the following solutions ultimately presented themselves:

(1) Certain building improvements were approved by the Ministry, but at that time toilet paper was still considered to be an unnecessary luxury. This is where the US Air Force stepped in. They wanted to facilitate reconstructing scientific research in Europe and were allocating funds. I was able to obtain some of these and could charge such trivialities as "special expenses".
(2) Using the money from the Air Force, I could hire scientists on a temporary basis and improve the salary of the low-paid assistants, but I was not able to create any permanent positions. The Ministry for Science agreed that a secretary was necessary for this amount of work, but the Ministry for Finances had implemented a strict hiring freeze. Then the Science Minister Dr. Heinrich Drimmel, recalled that he had a certain amount of resources to be used at his discretion, and he used this to finance my secretary.
(3) That was my job, and I began a lecture course dealing with all of theoretical physics, eight semesters long, five hours a week; and I offered a seminar about quantum mechanics.
(4) The Ford Foundation helped out with this by paying visiting scientists. I was only able to pull back the Iron Curtain a bit to our closest neighbors in Bratislava and Budapest over the course of time.
(5) Our experimental physicists did not consider the cooperation with CERN as true experimental physics; they were used to building their own experiments. I had no other choice but to establish a group in the department of theoretical physics which evaluated photo plates radiated in CERN.

I knew that these solutions were only temporary arrangements. The main thing missing was the courage to take on large projects. I hoped that the Austrian authorities would take over the financial support when I had shown that I was in a position to generate such results. This calculation paid off.

After these basic points had been cleared up, the most important minor point remained: my personal salary. I will summarize this long correspondence as a dialog.

Me: This is my paycheck, and I assume that with my increased responsibilities you will at least offer me as much as I had been earning.

Ministry: That won't work, because that's the top salary for professors. That would offend the older gentlemen if you were to receive so much from the very beginning.

Me: That might be true, but that is their problem. I don't see any reason to accept a decrease in salary.

Ministry: It is our principle that we hire younger people only at the first salary level.

Me: It goes against my principle to work more for less money. I know that I'm not the cheapest, but if you accept my conditions, I promise you that I will bring theoretical physics in Vienna to a world class level.

At this, the Ministry showed that this is what they were interested in and not in principles for principles' sake.

Ministry: I would suggest the following compromise: you will be hired at the first salary level, but I will pay you the difference to the highest level from another source. The consequence of this, of course, is that you will never get a raise.

Me: Later I will not have to work so much, and then there's no reason for me to get a raise. I'm assuming that the salaries here are adjusted for inflation.

Ministry: But of course.

I flatter myself that I fulfilled my promise and up to the very last years the Austrian state fulfilled its part of the bargain.

Master Years (1959 and After)

Vienna (1959–1968)

After my return to Vienna, Providence was smiling down at me and the individual program points were carried out as I had hoped; I just had to be patient for some of them to come to fruition. It was necessary to add on a fifth floor in the physics building in the Boltzmanngasse 5 to permanently solve the building situation, and this would end up taking 12 years. It took three tries to find the right secretary, but then she stayed with me until my retirement.

I have her to thank for being able to direct the constantly growing Institute without swelling up the administrative personnel. Franzi Wagner understood exactly what I wanted and she did it without me having to tell her. She was the true director. With the economic recovery, the number of employees at the Institute grew and there were visible improvements in the personnel resources at the University.

My lectures were at first unusual for many people, if not unintelligible. However, many were also able to adapt to meet the new challenges and I was able to train several first class physicists.

My fourth program point was more of a long-term project, which finally culminated in 1993 with the founding of the Erwin Schrödinger Institute. At first, the International Atomic Energy Agency paid for a visiting professor each year. However, this was just a drop in the bucket; there are currently 120 visiting professors each year at the Schrödinger Institute. I could see that people liked coming to Vienna; I simply needed to provide a pleasant working atmosphere at the Institute. The thing to avoid at all costs was having the few people who were there form even smaller cliques, with each believing that they are better than the others and being hypercritical of what the others were doing, observing it with jealousy and a sense of superiority. Group seminars, lunches, excursions and so on are the only way to counteract this kind of development.

While the contact to Western scientists remained purely a question of money, the exchange with the East was prevented by politics. In Hungary; for example, Georg Marx made great efforts to establish joint cooperation, but unexpected difficulties arose. Once he organized a conference at Lake Balaton and was even able to arrange the use of a guesthouse for party officials. Naturally, this was comfortably furnished and located in beautiful surroundings. I didn't want to miss this opportunity and drove there with my wife. The drive through the Puszta, the traditional Hungarian landscape, is very impressive, and at first everything looked like it would be a fantastic trip. But then the harm was done: my wife hurt her shoulder blade in a sudden careless movement and was in great pain. Our lodgings were so remote that there was no x-ray machine around far and wide, and the local doctor was unable to make a correct diagnosis. After two sleepless and painful nights, I realized that there was nothing else we could do; we had to return to Vienna and go to the hospital. What I had not counted on was that right then the wall was being built in Berlin, and Hungary was still a country under Soviet occupation. During the drive in, we had hardly noticed anything unusual, but now everyone was on high alert. The country was suddenly permeated with columns of tanks and all of the roads to the West were blocked off. We kept on having to turn back and after we had finally had our fill of looking into the barrels of machine guns, we returned to the conference venue without having achieved anything. After another sleepless night we tried again. Luckily there was now no sign of the tanks other than the tracks they left behind. I was finally able to bring my wife to the hospital and, after a few months, her shoulder was fine again.

One November morning in 1959, a young man came to me in my office after a lecture and said, "My name is Peter Freund, I've just arrived with a transport from Temesvar and am supposed to be deported to Israel the day after tomorrow. But I don't want to go, I'd like to stay here and get my doctorate under you. Could you arrange for a residence permit for me?" Fritz Houtermans had taught me that many Jews from the East are quite talented people and he would be a boon for our team. So I promised him that I would do my best. In fact, I was able to do what seems impossible today: he had a residence permit the very next day. My decision at that time paid off much later. Peter Freund not only received his doctorate at our Institute, but was eventually also offered a professorship in Chicago and became successful and wealthy. His bond with his old home country was preserved by his financing of the Schrödinger lectures in Temesvar. Western physicists continue to attend and present their results, so that Temesvar has remained connected to international scientific developments.

Freund was just the precursor; Romania was selling its Jews to the West for a few thousand dollars apiece. We helped wherever we could, and in 1968 we achieved something rather spectacular. A Romanian physicist had escaped via Czechoslovakia and landed exactly between the invading armies of the Warsaw Pact countries. The case was complicated by the following facts, which slowly emerged over time:

(1) He didn't have an Austrian visa.
(2) His wife didn't have an Austrian visa.
(3) His wife turned out not to be his wife, after all.

Despite these additional difficulties, we were able to bring both of them safely to Vienna.

These stories were spread around and were to my advantage when people started making accusations about anti-Semitism in Austria. I was deeply upset by them, as I had always been meticulous in ensuring that the question of social class, race or religion would not play a role at the Institute. One example will serve as an illustration: once a student came to me and complained that he failed an exam because the examiner was an anti-Semite. Naturally, I looked into this and discovered that the examining assistant himself was Jewish. No one knew this just because this wasn't the kind of thing people asked at the Institute.

While it was possible to maintain scientific contact with Hungary, although the contact was kept on the back burner, Czechoslovakia remained completely isolated until 1967. This had all the greater of an impact on me, because I could still remember how we used to take the streetcar to Pressburg/Bratislava before the war and buy shoes at the famous Bata shoe store. During group excursions with colleagues near Hainburg, located right across the Danube from Bratislava, we would look at the hills on the other side and ask ourselves what was going on over there. Then in 1967, I saw an announcement for a conference on superconductivity being organized by the University of Bratislava which did not say that Western participants were not allowed. As I was working on superconductivity at the time, I had the rather daring idea to simply register to attend. I remained the only Western participant and my impertinence was honored years later with an honorary doctorate from the Comenius University in Bratislava.

The people from Bratislava were ecstatic to meet their neighbors and wanted to work together. I said that it was just a little over an hour's drive, we could hold one seminar in Vienna and the next one be organized in Bratislava. We were able to carry out this plan and the absurdity of the border could already be seen in the participants' last names. From Bratislava came Mr. Kessler and a Mr. Mayernik, a slavified Mr. Mayer, to Vienna to meet with a Mr. Jelinek and Mr. Konetzny. The news of our cooperation spread, and some people from Budapest signaled their interest in participating. They were naturally more than welcome, and in this way what later would be a flourishing institution at the initiative of Herbert Pietschmann known as the Triangle Seminar was developed.

Of course, the Bratislava leg was cut off after the invasion of Warsaw Pact tanks in 1968, but the seminar had gained other sources of support and was able to survive. The peoples who had so enthusiastically celebrated their independence in 1918 were now pleased to be reunited through the reconstruction of the Danube monarchy at the scientific level.

I can finally turn to my fifth program point, joining CERN. In Berne, I had learned the technique of how to study high energy particles in a photo emulsion, my wife had worked there as a scanner. She had even discovered the eighth τ-meson in the world after searching for three months. This technique was relatively inexpensive, you just needed enough microscopes. The same technique was also being used by the new machine PS in CERN. It seemed the logical way to enter experimental high energy physics in Vienna. The method had been developed twenty years before by Marietta Blau and was based on the following: when a charged particle crosses a photo emulsion, it ionizes the atoms along its path and in the following development process, the silver spheres are blackened at the ions. Therefore, charged particles leave trails that were thinner the faster the particle flew and therefore had less time to ram into atoms. At around half the speed of light, Einstein came to the rescue and ionization once again increased. Through the contraction in length, the electric field is squashed out wide and is able to reach atoms located farther away.

If there is so much silver in an emulsion that the particles, even with this minimum of ionization, have enough blackened silver spheres to be visible, then you have the perfect detector; no charged particle can escape unnoticed. Cecile Powell from the University of Bristol had developed this kind of emulsion with the Illford company. When the large accelerator began functioning in 1946 in Berkeley, his co-worker Bepo Ochiallini was present and, in fact, the first artificially made π-meson could be seen on the plate.

Our experimental physicists considered these goings on not to be real experimental physics, but I thought it was magnificent. Here you could see with your own eyes all the atomic processes you usually only read about. You could see a thin trail, so it must be a fast particle! It had to be light, because it was easily thrown off its path, so it must be an electron. Another trail was a bit thicker and at the end began to wobble a bit, the particle must be slower and heavier, so it's a proton that was stopped. These phenomena formed the less interesting background for our actual task at hand, it was certainly more exciting when two trails began at one point and then got thicker. An uncharged particle decayed into a positive and a negative, the thickness of the trail didn't depend on the sign of the charge, but solely from the speed. You could see even more in the bubble chamber pictures, because there you didn't just see a tiny section, but what was happening over a length of two meters. A magnetic field was also placed in the chamber, and that showed the charge sign and the mass of the particles.

I can't help but present the reader with a picture where you can immediately see many elementary particles.

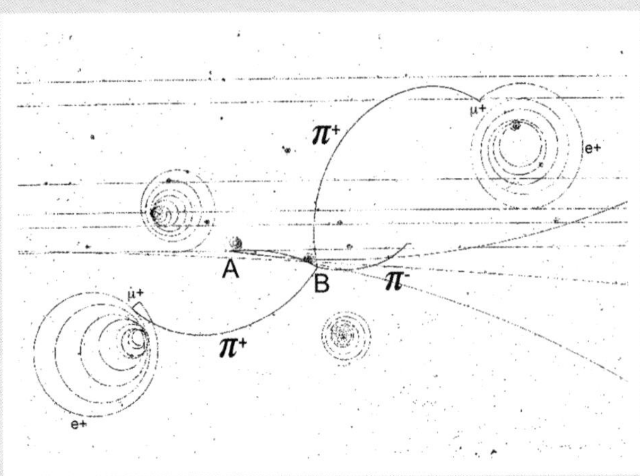

Fig. VI: Meson decay

Let me begin by describing how a charged particle moves in a magnetic field. It feels a force perpendicular to the magnetic field and to its speed (the Lorentz force). It therefore runs in a circle which is perpendicular to the magnetic field, and the following holds true:

Orbital period = radius/velocity = mass/charge times field strength

The frequency is therefore not dependent on speed; for faster particles the orbit has a greater radius. If you want to keep the radius constant during acceleration, you will need to increase the magnetic field in proportion to the speed. In our picture, the magnetic field is, of course, constant, and the faster and heavier particles have a larger orbital radius. The Lorentz force changes its sign with the charge, so that the orbits of oppositely charged particles are curved in opposite directions.

We can see a lot of mesons decaying in the picture. A K+ is produced at position A and decays at B to $2\pi^+ + \pi^-$*, a so-called* τ *decay. (When my wife discovered the eighth* τ*- meson it was still thought to have been a new particle, but it is actually a type of decay from the*

K-mesons). The signs for the charges can be seen in how the orbits are curved. The negatives are curved like small spirals which are the orbits from electrons which have been punched out of their atoms. Now the π^+ decay into τ^+ + neutrino (invisible), while the π^- is pulled to the atomic nuclei and ultimately absorbed by it. The τ^+ then decay again into e^+ + neutrino- anti-neutrino (invisible). e^+ (positrons) are as light as electrons, and therefore their orbits have a smaller radius. In just one glance, this picture shows you all of the particles' fates.

For me, this was just a sideline, but luckily Marietta Blau returned towards the end of her life to live out her days in Vienna. She was the real inventor of the emulsion technique but she never had the possibility to bring it to fruition. She never received the recognition that was her due, and also in Vienna her modesty prevented her from beating her own drum. I had originally hoped to win her to be the director of the whole group, but she didn't want to take on any responsibilities that she would not be able to fulfill. She did, however, make herself available to advise PhD candidates on a volunteer basis.

Naturally, she was appalled by our meager equipment, as she was used to something else entirely in the USA. I could only tell her that you had to start small. The group flourished thanks to the competent and motivated students who had quickly learned what to do. They were led by Brigitte Buschbeck and several of them including Wolfram Kittel, Gottfried Kellner, Ingrid Dahl-Jensen, Hannelore Sexl, Hermann Jeram, Gerhart Capek, Horst Wahl, and Gerd Otter were offered leading positions all over the world.

Still, this was just the beginning, the emulsion technique was exhausted after a certain period of time. It was replaced by the bubble chamber technique which soon presented greater challenges. But this technique could not deal with large amounts of data, either. The times when we were spending months looking for a certain decay of a K-meson were over; the new machines could produce millions a second. But how was I supposed to keep up? Our Institute was not in a position to cope with these challenges; we had neither the space nor the resources. In this situation, Professor Fritz Regler from the Technical University came to the rescue. He was a member of the Austrian Academy of Sciences and we proposed that the Academy transform itself from a debating society into a research institution. This required a bit of persuasion, but the initiative was ultimately successful, and soon the Institute for High Energy Physics of the Austrian Academy of Sciences was established.

Our "plate group" became a part of this institute and in this way we were able to continue our work in physics as it should be. The institute's directors were theoretical physicists beginning with Wolfgang Kummer, then Herbert Pietschmann and finally Walter Maierotto. They knew how to get the institute included in various international projects. After all, it was also involved in the UA1 experiment at CERN, in which the W and Z particles were discovered. This discovery earned Carlo Rubbia, who led the experiment, the Nobel Prize. At that point in time Europe was already superior to the USA in terms of equipment, and the disdain that had been shown to our resources was a thing of the past.

Meanwhile, CERN was unflinchingly following its path to become one of the best accelerator labs in the world. It had already constructed two machines; the small SC (supercyclotron) and the large PS (proton synchrotron) which were already functioning perfectly (and continue to do so today). CERN's engineers, led by John Adams, had demonstrated their dependability, as everything was carried out exactly on schedule and within budget. Now a crossroads had been reached, as the particles in the PS were being accelerated almost at the speed of light. As previously mentioned, the energy available for particle production increases at further acceleration only with the root of the primary energy. Most energy is placed on the particle hit, and there is only a little energy left for the luxury of producing new particles. So when I want to produce a particle with 100 GeV (giga electron volts), I first need a ray with $(100)^2$ GeV = 10.000 GeV energy. That's one expensive hobby! The only way to go farther was pursued by Bruno Touschek with the colliding beam technique he developed, in which two beams are shot at each other. If they run into each other and hit each other on the head, both adversaries can calm down and all of the invested energy is now available for the production of particles. Of course, it's not possible to really aim to get a direct hit, but you can let the particles run into each other again and again in a circle, until they just happen to hit each other just right. Of course, this requires new technology so that the aim can at least be good enough so that the particles never reach the walls of the vacuum tube. After all, in 15 minutes they almost travel the distance of going from the Earth to the Sun and back twice. The only place in Europe where it was possible to go on this kind of adventure was CERN. These Intersecting Storage Rings (ISR) presented such a large undertaking that their construction required its own resolution from the CERN Council. The project was not without controversy, some physicists thought that the smallest wavelength had already been reached with the energy of a GeV. Higher energies that would produce smaller wavelengths would not serve a purpose, as there would be nothing to see. I did not share this opinion because the energy of 1 GeV was only distinguished by the mass of a proton, which seemed to me to be more of a coincidental circumstance. The more significant thing for me was the interaction strengths of weak interaction which define a level of energy of about 100 GeV. So the little window we need to look through to see the actual kitchen where the fundamental particles are being brewed is way up there. I could think of no better ladder to climb to reach that window than ISR technology.

At the time of the proposal of the first ISR, I was the Austrian representative in the CERN Council, and Fritz Regler was my deputy. It was up to us to be sufficiently persuasive in Austria so that Austria would officially take a positive position. But there was more at stake. There had to be a unanimous vote in the CERN Council, meaning that in all the other member states, the governments had to be simultaneously sufficiently well-disposed toward science so that they would agree. Years of negotiating were necessary until the magic moment arrived. Then a special Council meeting was called, and when we arrived in Geneva, there was a euphoric atmosphere amongst the delegates. Everyone was aware of the significance of the moment.

One hour before the vote the bomb was dropped. I received instructions from Vienna to vote against the ISR. This was simply a catastrophe, as there would hardly be another opportunity this good. Ironically, right now the Austrian proportional system, of all things, could help me out. Although I had never been a member of a political party, I was considered to be a liberal, and therefore, the conservative Regler had been appointed as my deputy. At that time, the Minister for Science was a conservative, and Fritz Regler knew what he had to do. He simply said that, "the Minister is in the same party, I'll be able to explain to him why right now Austria can't block the vote. This time we'll vote against our instructions and I'll settle things in Vienna."

I put my faith in this backroom deal and that it would work, and voted in favor of ISR. I was not suspended for disobedience, and the path was cleared for what would become a triumphal procession for European science.

I had had little time for scientific research during my first years in Vienna, but I had been able to work a bit ahead and succeeded in staying abreast of developments. For example, I was familiar with the trains of thought that would lead to today's standard model of elementary particles, and could make a few contributions which finally resulted in my being offered a position in the board of directors at CERN.

Fig. 16: CERN

Geneva (1968–1971)

When you travel to Geneva by airplane today, you can see a settlement of rather low, long stretched-out buildings that are clustered around a tall central building, located between Lake Geneva and the Jura mountains. You would never guess that this contains the largest particle accelerator in the world. The accelerator itself with its length of 27 km rests between 50 and 100 meters under the fields and mountains. There are entrance shafts found at only four places; these have a typical dome-like structure. This massive instrument is supported by 21 European countries and several thousand people are working here, including, of course, a lot of scientists, but mostly engineers and all kinds of service personnel.

The construction of such a monster project requires many years, of course. The planning and the decision to go ahead with this development took place exactly during the time when I was a director at CERN. There were also two experimental physicists, two engineers and an administrator with me in the Directorate. The Director-General was Bernhard Gregory, who was French. Around this time it had become clear, which was then later proven to be true, that progress in elementary particle physics was only possible when higher energies would be available. For this, larger accelerators were necessary, which were naturally ever more expensive. For most European countries this kind of financial expenditure for a single branch of pure science was simply not reasonable. The solution was called "globalization".

The question was, how far did we want to go? The logical solution would be a "world machine", and this was often brought up in discussions. However, we were still in the era of the Cold War, and there were doubts as to how well that would work. Under no circumstances could anyone afford a failure, so the idea of a world machine was laid to rest. As CERN was working out so well, the tendency was to go with a European solution, but then the question was where should the machine be built?

All of the countries recognized the stimulating effect CERN had on the surrounding area. Not only were there lots of high-paying jobs for the people in the area, but also many high-tech companies settled or were founded near Geneva. People wanted to be on the scene when commissions were assigned. Therefore, every country wanted to have this accelerator on its territory. Many proposals were made, with each country wanting to present itself as being more attractive than the next. The CERN Directorate and representatives of other countries were all invited to inspect all of the proposals and were flattered accordingly. However, the quality of the food and wine doesn't play such an important role when it comes to deciding on the location. The main things to take into consideration are the quality of the soil where the machine will be, the traffic situation and transportation infrastructure, as well as the living conditions for the international personnel.

Austria put forth Göpfritz in the Waldviertel (Forest Quarter) area, highlighting its advantage of having granite bedrock. Naturally, this area is not exactly a suburb of Vienna, and to make the journey seem shorter, we were flown in by helicopter. However, on the trip back after dinner in the Wachau valley, the real distance could not be denied. There were advantages and disadvantages for every proposed location. France had found a beautiful spot in the Provence, and although the wine was undeniably good, it was somewhat remote. Germany suggested a place in the area near Münster, but even referring to the historical role that this area had played in the Peace of Westphalia couldn't prevent the spoiled people from CERN from remarking that they would become depressed in such a monotonously flat landscape. There were also several things to criticize with the proposals from Belgium and Italy. There were the many questions to be answered about the geological formation of the locations. The granite in the Austrian Forest Quarter showed some cracks, which disqualified Göpfritz.

We knew that we were spoilt for choice, but still we held out hope that the politicians would somehow be able to arrive at an agreement. However, here we were mistaken; the politicians did not move an inch. Every country's representative insisted that only their proposed location be accepted, otherwise the project would fail. What was to be done? Should we just give up?

Bernhard Gregory came up with a plan to get us out of our dilemma. He had realized that national egoism would lead to nothing and although he was a member of the French upper class, he dropped the French proposal. He even raised the question in the Directorate if a second European lab would even make sense, considering the mentalities of the politicians. We could only confirm his doubts. We had already experienced so much haggling about for the lab in Switzerland that we could vividly imagine what would happen if the funding available would have to be split between two labs. Therefore, we abandoned this plan, and history has shown that this was the right decision.

Later on, the Americans had the exact same problem when they were looking for a place to build the somewhat even more ambitious accelerator SSC. The Texas die-hards were able to prevail and a new lab was built somewhere in Texas — in the middle of nowhere, which devoured an enormous amount of money. After the appropriated 2.5 billion dollars had been spent, it was concluded that the final cost of building the machine would amount to 10 billion dollars. Here, the politicians lost their patience and said enough is enough, there will be no more funding. So the physicists were left standing there with a giant tunnel without an accelerator, which was an enormous setback for American particle physics.

But what could we do in our situation? The only logical alternatives were either to drop the project entirely or to build it near CERN and to sell it to the politicians as an enlargement of CERN. Of course, a new project requires new planning, but no one should know about it, because if a country were to find out that its proposal had been rejected, it was sure to be offended and would oppose the new project. Gregory therefore demanded that we keep absolutely silent about it, which we were able to uphold for months. My friends at CERN were upset with me for having lied to them for so long, but I couldn't risk endangering Gregory's plan in any way.

Now we were faced with the problem of making this complete change in course believable. Why hadn't Geneva been considered in the first place? The reason was the soil, the moraine from the Ice Age glaciers that everything was now standing on. The question is: which consistency is necessary for such a machine? This is a purely technical question and can only be answered by the experts. Our machine technicians simply said that recent studies had shown that the soil conditions would allow construction near CERN. The politicians accepted it, but were we telling the truth? Yes and no. For the machine that was being developed at the time, the SPS, the answer is yes. For it, there were no problems concerning the consistency of the soil. Later; however, much more serious and unexpected difficulties arose, such as while building the tunnel under the Jura, suddenly coming upon a natural torrent which immediately flooded the entire construction site. The recent further addition of the LHC (Large Hadron Collider) is so gigantic that the consistency of the soil could be problematic. However, back then we weren't even daring to look this far into the future.

We could claim that the new idea would be the more cost effective alternative. Not only was the infrastructure already in place, but the old PS (proton synchrotron) could be reused as an injector for the new machine. This was then called the super proton synchrotron (SPS) to make it clear that it was an enlargement of the PS. Of course, this was something of a white lie, because these kinds of machines have to be constantly renewed, and there was hardly anything left over from the old PS, but it sounded good.

In any case, we had to deliver a long-term cost analysis for the new project, which went something like this: we began with the principle that the Western industrialized nations should allocate a minimum of 2% of the national product for research and development. Our reason for this was that successful industries took even higher percentages. From this sum, 10% was to be reserved for pure basic research. Taking these numbers, the expenses for particle physics caused by the construction of the SPS would amount to 10% of the money spent on basic research. The objection that many other branches of science had to be funded with this money was met by pointing out that we had calculated our figures using the GNPs as they were at that time. However, we were assuming that the national products would increase in real terms, and then there would be more room for other sciences which also required such huge facilities. We also mentioned that wealthier nations could afford more luxury, and the 2% could also be increased. All of this turned out to be true, so that our arguments at the time were not just hot air. The thing we had not taken into consideration is that the number of member states in CERN would double and the financial burden could be spread thinner.

Now came the most important question. After having planned such a giant machine, was there even a person who would be able to realize such a monster? Gregory had thought of this as well, and enlisted John Adams. He was a self-educated Englishman who turned out to be a technical genius. He was for accelerators what Wernher von Braun was for aeronautics. As long has he lived, everything worked. He had already built the PS and we were fully confident that he would be just as successful with the SPS. In fact, we had bet on the right horse: the SPS was completed on time and within its budget. This accomplishment was a triumph for European technology. Unfortunately, this standard could not be upheld during the most recent expansion, the LHC.

Fig. 17: John Adams

The activities described thus far at CERN were not work in the usual sense; I just had to keep a cool head and not lose sight of the big picture. The work was also sweetened by the magnificent location of the Directorate. Our offices were in the top floor of the large administration building. There was nothing facing it and in clear weather there was the Mont Blanc massif for a backdrop glowing through the window. I never got tired of the view, but my colleagues, somehow infected with the ascetic spirit of Calvin, blocked it out with venetian blinds.

My real job was being the Director of the Theoretical Division, and this was made easy for me by my excellent secretary, Helga Schmal. The Division was rather small, of the almost 100 people working there only 10 were permanent employees at CERN. The rest were temporary staff. Whenever I urged to enlarge the group, I always received the answer that the theoreticians were actually unnecessary here, as they were only dealing with such abstract and unreal things that were useless in experiments. Still, the group of permanently employed researchers were the best brains in the business from their countries.

There were Jaques Prentki (elementary particle physics) and Andrè Martin (mathematical physics) from France, Rolf Hagedorn (quantum statistics) from Germany, Thorleif Ericson (nuclear physics) from Sweden, Lèon von Hove (statistical and quantum physics) from Belgium, Sergio Fubini (elementary particle physics) and Jurko Glaser (mathematical physics) from Italy, and John Bell (universalist) from Ireland. With this kind of a concentration of expertise in one group, it was somewhat delicate for me to call their working methods into question. At the end of the day, I was able to find some people who were willing to work with experimental data directly. This did not garner me much praise, rather the reactions were more dismissive. It was said that we just wanted to have the measurement data handed to us so that we could use it as a basis, work with it and receive all the credit for it; but the others could do the same thing for themselves, actually even better. I gave up trying to justify our existence and worked as I saw fit.

This attitude expressed by the experimental physicists influenced my later decision when I was able to obtain one new permanent position in the theoretical Division in three years. As it was well paid and made no further demands on the candidate, there were naturally many applicants for the post. There was also the fact that no nation was to feel that they were at a disadvantage, and therefore I could not prefer an Austrian, even though Julius Wess seemed to me to be the most qualified candidate. The long list of applicants contained primarily people who had become attached to current trends.

As mentioned above, the basic disciplines in the theory group at CERN were already optimally filled, and we were looking for a new area relevant for high energy physics. It therefore seemed to me to be more important to find an independent thinker, even when their results could not be immediately used. I finally decided on Bruno Zumino, even though we already had a surplus of Italians. I had already gotten to know him in Göttingen. He had been in Vienna for a time, and strangely enough had not been offered a prestigious position in Italy, so he emigrated to the USA. Now it seemed that CERN could offer the possibility to counteract the European "brain drain".

I was at least partially successful. He accepted the offer from CERN, and together with Julius Wess, introduced one of the most productive working areas in theoretical physics in the last century, supersymmetry. Later, he returned to the USA for personal reasons.

When I came to CERN, the evidence for the quark model was so convincing, that for me, the problem of elementary particles was mainly solved. I soon realized that many others were still following completely different concepts. This was a little frustrating for me to see, because a lot of it seemed useless, but in my position I couldn't just suppress different tendencies. I switched over to concentrating on completely different problems and that's how the following escapade came to be.

At a conference I heard a lecture by the world famous English astronomer Fred Hoyle. He was an original thinker who didn't shy away from breaking taboos. This meant that he was often considered to be an outsider, and in some quarters was denounced as a fool; even when he said something that was obviously correct, it was rejected. He had predicted the resonance of the ^{12}C nucleus, because nuclear synthesis in stars functions through them, and we would not exist without carbon. No one wanted to accept the fact that nature had arranged something just for our sakes. They were completely forgetting the fact that we do, in fact, exist. Then, of course, the measurements from William Fowler showed that this resonance is present. Fowler even won the Nobel Prize for this discovery, while Hoyle remained empty-handed.

At this conference, Hoyle had another provocation ready, he praised stars for being such good heat generators. He seemed to be saying that they got hotter when they radiated energy. This made me very confused. It would mean that they had a negative specific heat. Specific heat measures the change in temperature with the change in energy and in my lectures I had always been able to present it as a variable square of the energy. The square of a real number is always positive; how can it become negative? Something must be wrong here. At first I thought to myself, this is not my area, they should make their own mistakes, I should keep out of it. But then it went against my sense of honor to just walk away; after all, I had proven that just the opposite was true. I began thinking things over and remembered that the proof of positivity for specific heats was only valid in a system that was in equilibrium with a heat bath. This was only possible for a system with positive specific heat. If it was negative and the system were to give up energy to the heat bath due to a fluctuation, then the system would warm itself up again and even more energy would flow into the heat bath. The exchange of energy would be built up and as long as the specific heat was negative, it would not be still.

A system of negative specific heat must, therefore, be isolated, and its energy must be constant. The question was just where would it be possible to cultivate such a quirky thing? And do such things even exist? To find out, I began tinkering with different mathematical models and thought that it had to be something very ingenious since nothing like it had ever been found before. However, I was to find success only after I played dumb and said to myself that gravitational interaction is something like a potential hole, but each particle attracts every other one. So I made the hole N times deeper when N particles found themselves in the hole. This hole in the bucket trick was so simple that I was able to calculate the thermal properties exactly which showed that the model did indeed have a negative specific heat.

I was thrilled. I wrote it all down and handed it in for publication. My enthusiasm did not carry over to the journal's editor, who rejected my findings. Not that he said that my bucket game was too trivial, or that the astronomers knew this already. Rather, it was just the opposite: he said that this was too revolutionary; something had to be wrong with it. He was a true gentleman and didn't want to simply shake me off, but invited me to give a lecture. He had invited all of the experts in statistical physics from the surrounding areas and they were to urge me to foreswear my false beliefs. There I stood like Martin Luther before Emperor Charles V and could only say that I cannot recant anything against my conscience and knowledge. Only I didn't have to say, "God help me, amen," and I didn't have to flee to Wartburg afterwards. Still, the editor was intent on quelling this heresy. When no one was able to find a mistake in my math he said, "If what they say is true, then the conditions of statistical mechanics have not been met, and noting guarantees that the properties you have found have anything to do with the dynamics of the system. I find the whole thing to be nonsense, but if you want to make a fool of yourself, then you can publish it." I took the risk and had it published.

In the following years my collaborators and I were able to make the system more true to life and there were dozens of examples of systems with negative specific heat in the literature in the meantime. However, the editor's objection could not be refuted for a long time, as who was able to calculate the movements of so many mutually attractive particles? Twenty-five years later the tide had turned, and the computer began its era of triumph. The movement of hundreds of mutually attractive particles is calculated in a matter of seconds by today's computers. Three Dutchmen made the calculations for the case I laid out. Their results confirmed my predictions perfectly. The particles made a hole for themselves, which most of them would fall into. The lower that the energy became from that, the more particles would snuggle up to each other and therefore the hotter it would become. Had the editor made a fool of himself? Not entirely, the computer showed something which had eluded my thermodynamic reflections. Namely, that at first it gets much hotter in the hole with the particles than it is outside, and only much later do the temperatures adjust to one another. When we interpret the clumping together of the particles as we would for the creation of a star, then we're hot on the trail for the basis of life. This is sunshine, and this exists only because it's much hotter on the sun than it is here by us. This is only possible because the balance when all temperature differences have adjusted and evened out has not yet been reached. Therefore, the most important thing for us is thermodynamic instability, according to which the original homogenous temperature and mass distribution is not stable, but rather through the building of clumps, differences in temperature are created.

Our years in Geneva were the best years of our life in terms of quality of life. We didn't live in Geneva itself, but in Nyon, where we had rented a house from the actress Liselotte Pulver that was located directly on Lake Geneva. We enjoyed our own private beach which we had to share only with the occasional wild swan. From our living room Mont Blanc offered a magnificent background overlooking the lake. Of course, all this luxury came at the price of having to do a fair amount of driving; I had to get to CERN and my wife took both of our sons to the international school. Everything worked out well; my wife and her Mini only came into conflict with the police once for speeding, much to my sons' amusement. My older son was going through that phase of wanting to be a race car driver and classified my wife's driving style as snail-like. My wife had to drive fast because she had a lot to do; there were many social commitments connected to my position. She ran our home without any outside help and used the money for a maid to support her aunts in Vienna. Although they were refugees themselves, they had been the ones who had scrimped and saved to enable her to study after the war.

My wife was rewarded for all of her efforts by many interesting friendships and acquaintances that we established. However, for our children the move was quite difficult at the beginning. This was a different country, a different school and with two different languages. Our older son eventually transferred his racing ambitions from the street to the ski slopes and the magnificent possibilities offered around Geneva soon completely reconciled him to his fate. It took longer for our younger son, and this could also be seen in his schoolwork. When my wife asked him what she could do so that he would no longer be so unhappy, he replied that there was something. If she would allow him to play soccer for the FC Nyon team, he would also do better in school. My wife agreed, saying that if his report card had improved by Christmas, then he could play. His grades got better and the soccer team soon became his life. He had quickly become completely integrated and was at home on all of the soccer fields in western Switzerland. When we left three years later to return to Vienna, his leaving was regretted by the local press (but not mine). We also have this physical activity to thank for the fact that our sons proved to be immune against the affluent squalor that was rampant in the local schools. One of our younger son's friends always paid with a 100 frank note, he never had anything smaller, although his father had already been deposed as King of Libya by Ghadaffi. There were dangers to the kids lurking everywhere: stealing, smoking, drugs. One of my colleagues had lost both of his children to drugs. His daughter committed suicide and his son had wound up in a clinic for incurable addicts. How happy we were that our sons didn't even want to smoke cigarettes.

When Gregory offered me a permanent post at CERN, I was sorely tempted. That I ultimately decided to return to Vienna cannot perhaps be completely understood on a rational level, but there were several things in favor of doing so. I didn't have any actual task in the Directorate, but was there to lend a voice of reason. Of course, although this was important, it would have been temporary. As a researcher, I would have been required to invest all of my energy into elementary particle physics, which I couldn't guarantee with my wandering interests. I also saw how in some countries, for example, at that time in Italy still, my colleagues were being paid poorly or not at all. But these countries were still responsible for financing CERN and therefore our salaries. It was scandalous to me that I would be partially financed by these countries so that I could follow my hermetic interests in luxury, or even just play the organ. In Vienna, I could at least earn my keep with my lectures.

Maybe I also just had the vague notion of not having completely fulfilled the pledges I had made when I was appointed. I could not yet have an inkling that I would achieve all of this with the founding of the Erwin Schrödinger Institute. In any case I used Gregory's offer to pressure the Ministry in Vienna to finally finish the fifth story at the Institute. The enlargement of our Institute and the constant expansion of the library had made the building situation unbearable. At that time, the fifth story was criticized for being wasteful, today the millions spent are forgotten and it's impossible to imagine how things could have worked without it.

Return to Vienna (1971)

When I returned to Vienna, I had reached an age where many scientists consider their creative phase to be over and they begin looking for a position on which to rest their laurels such as decan, dean, or president of something. This would not have satisfied my cravings; for me scientific research was like a drug — my thoughts could not rest in the face of an unsolved problem. However, it can be risky to conduct research alone, particularly when it's contaminated with the obstinacy of old age. Many great scientists had maneuvered themselves into obscurity this way, and I owe the fact that I was able to avoid this to the people who accompanied me along the way. Unfortunately, I lost two of my most talented colleagues, Roman Sexl and Alfred Wehrl, through their untimely early deaths. Fortunately, there were always excellent researchers who came after them, such as Manfred Breitenecker, Harald Grosse, and Bernhard Baumgartner; an inexhaustible series of great talent. I mainly worked together with three people who had entirely different characters: Elliott Lieb, Heide Narnhofer and Harald Posch.

In an intense collaboration, ideas spring out of discussions and thoughts melt together as one. It can be the case that the light bulb suddenly goes on and the results are immediately present, or it can be that years of work are required for the solution. The question of who achieved which result in our work never came up; therefore, I always made sure that our names were listed in alphabetical order. In the next part of the book I will try to guide the reader into the mental world which so fascinated my colleagues and myself. I was always interested in fundamentals and never considered financially profitable applications. Feynman put it best when he said, "Physics is like sex. It may have consequences, but that is not the reason why we do it."

Stability of Matter

The matter around us is constantly changing its form, but the entire quantity remains the same throughout all of these reshapings. For the

Greek philosopher Democritus, matter consists of tiny, indestructible parts, the atoms. The different compositions of the atoms are the reasons for the variety in the form of matter. However, the atoms were so small that it would take over two thousand years before anyone was able to find them. And then they were not how we had imagined them to be. Their electron shell could easily be torn open. The much smaller and heavier nucleus appeared to be more robust, although certain minor adjustments were observed in radioactivity. The fixed stars appeared to possess the other kind of immortality. The fixed stars are so powerful that it was considered to be impossible that they could one day be destroyed. By the middle of the 20th century it became clear that this stability, whether very small or very large, was deceptive. Atomic nuclei burst in the middle when they get too fat, and when stars get too fat, they collapse onto themselves.

These discoveries had been predicted by researchers years before they became common knowledge, but no one had wanted to listen. Even the most important experts proved to be blind in this regard. In the early 1930's, Enrico Fermi discovered radioactivity while radiating uranium with neutrons, which he interpreted as transuranium. In 1934, the German chemist Ida Noddack published an article in which she clearly stated that this result was not definitive, because Fermi could not eliminate the possibility that the uranium nuclei had split into two more or less equal parts. This is exactly what had happened, but no one wanted to see it, not Fermi, not Einstein nor Heisenberg et al. Only four years later was this fission product proven chemically by Hahn and Strassmann. This method was like taking a detour, because so much energy is created during fission that it's easy to see directly. After its discovery, it was observed at different locations within days. A colleague of mine told me that they had long seen flashes when the spontaneous fission had occurred, but it happened so rarely that they just attributed it to equipment failure.

The story of gravitational collapse is similar, only it took place over a longer period of time. Two years after the laws of quantum mechanics had been established, in 1928, Isidor Frenkel mentioned in an article about metal electrons that a sufficiently large body, when continually shrunken by gravity, would release more and more energy until it had contracted into a point. This result was rediscovered by several different researchers and carried out further, most impressively in 1930 by a young Indian named Subrahmanyan Chandrasekhar. Even though it was a necessary result required by the laws of nature as they were known at the time, no one wanted to believe it. This went so far that the famous astronomer Eddington cancelled a scheduled lecture by Chandrasekhar at a conference in order to avoid a public scandal. He was not even allowed to comment on Eddington's lecture. Thirty more years were to pass before this phenomenon became widely accepted as true and Chandrasekhar was to receive the Nobel Prize. Since then, physicists have become more sensitive towards the stability of matter.

Basically it goes back to the question that Goethe posed in "Faust", namely what holds the world together inside? By now, physics should be able to provide an answer: the world around us consists of electrons and nuclei and the forces between them are known. These are primarily the electrostatic repulsion between the electrons and between the atomic nuclei, while the electrons and the nuclei attract one another electrostatically. But does this hold everything together? Does repulsion or attraction win out? Or, to put it in more general terms, are things exploding or imploding?

When we allow the forces in Newtonian mechanics to work, then the answer is neither nor — things are crumbling. This is because the electrostatic attraction between two charges can become as strong as it needs the closer they get, and already with three particles they get so close together that they have enough energy to throw the third one out. This means that groups of more than two particles do not stay together. They always dissolve into pairs of oppositely charged particles. We can see this in the sky; almost half of all stars are double stars.

It is a clever move in quantum mechanics that "prevents" this disintegration and makes it possible for the unbelievable abundance of molecules and larger bodies. Quantum mechanics can easily answer Faust's question and eliminate the possibility of an explosion. Heisenberg's uncertainty relation does not allow for a particle to be more precisely localized in a phase space than in a volume the size of Planck's constant. The phase space is the space of the location and the impulse and Heisenberg proved that when you want to limit one of these quantities, the other one needs even more leeway. This also means that you cannot press two particles together too much without having them fling about wildly, which would consume the electrostatic attraction energy. Not enough energy is won through the pressing together of two to challenge the rest of the particles to come out; the system does not disintegrate.

What saves us from the dark specter of implosion? What prevents matter from disintegrating? Physicists were not aware of this danger for a long time, but it is real. When many particles stick together and act collectively, then the increase in kinetic energy mandated by Heisenberg will have less of an impact than the electrostatic energy won. This will make matter clump together even more, the more particles that are involved. Quantum theory by itself cannot prevent this clumping. It takes place among the so-called bosons. Incredible amounts of energy would be released and it would come to a cosmic catastrophe if long-lived elementary charged bosons existed. We have to thank the so-called Pauli principle, a special property of electrons, for the fact that this does not happen.

Actually, this danger should have been clear already in 1926, after the laws of quantum mechanics had been discovered. But it took another 40 years until Freeman Dyson and Andrew Lenard could prove that Pauli had, in fact, prevented the catastrophe with his principle and that the familiar particles could peacefully co-exist.

The Pauli principle is a continuation of the Heisenberg principle for many particles. It can be defined with mathematical precision using symmetry properties of the Schrödinger function, but for common usage we can simply say that the electrons have illusions of grandeur and demand their own room in the phase space. Heisenberg demands that the room not be too small and Pauli says that no one else can already be living there. This successfully prevents the previously mentioned clumping together of matter.

Now the reader might be thinking, this is all well and good, but Dyson and Lenard have solved the problem. It's over, why should I bother with it anymore? Elliott Lieb and I thought differently and what we discovered in the course of our research opened up completely new areas in mathematical physics.

I would now like to say a few words about Elliott Lieb. I took part in a conference in Trieste in 1968 where a young American presented a short lecture about the residual entropy in ice. This meant that the condition of the lowest energy of ice did not show entropy of zero, but can be realized through many configurations. Showing these configurations is an extremely tricky thing to do mathematically. The two time Nobel Prize winner Linus Pauling had an idea of how, but couldn't prove it. Elliott was not only able to describe Pauling's idea at Trieste, he was able to present the solution of the problem. I found his lecture better than some of the main ones. He had clearly defined his starting position and didn't get off track or put a gloss on the situation, but made a solid series of deductions. This style had become rare in theoretical physics, and I was hoping that he would usher it back in.

My expectations were not to be disappointed, and two years later I found an article from Lieb and Joel Lebowitz, in which the authors proved that for systems that just have electrostatic interaction according to quantum mechanics, the thermodynamic functions possess the desired properties as long as all particles of one sign of charge keep with the Pauli principle. This means in particular that for such systems a negative specific heat could not appear, and this distinguishes the Coulomb systems from those gravitating systems I had been researching.

I was at CERN at that time and thought that this result belonged in the basic training of every physicist. So, I invited Elliott to give a lecture. When I picked him up at the airport, I noticed that he didn't have a briefcase with him and I asked him, a little surprised, if he'd forgotten his notes for the lecture. He just tapped on his forehead, where everything was stored.

His brilliant lecture, which he gave off the cuff, made a strong impression on the audience. His thought process was crystal clear, but his proof of thermal stability was based on Dyson and Lenard's proof of another stability previously mentioned. This says that the energy stored in any conglomerate of N particles could not grow stronger than in proportion to N. This means that two gallons of gas cannot have more than twice the energy of one. Dyson and Lenard's result sounded all well and good, there was just one catch: the proportionality constant. The proof resembled mathematical fireworks, forty pages with countless inequations, tricks and estimates. The length was not only detrimental to intelligibility, but also to precision. As the result was so general, the inequations were never sharp. If we were losing a factor of two on each page, though, this means after 40 pages a factor of 2^{40}, which is around 10^{14}. Therefore, the limit was valid for the stored energy for bodies of every size, but it was always a hundred thousand billion times greater than the actual energy.

With this in mind, I suggested to Elliott after his lecture that to make the whole structure more transparent and provide realistic figures, better proof for the stability of matter was needed. Elliott shared my view and we agreed to examine the problem. It would take years before we found the correct way, even though the situation is quite simple intuitively. To spare the reader from having to follow the argument in all of its twists and turns, I have banned the physics thinking behind our proof to a box. We assigned ourselves the task to solve the problem in its entirety and purely mathematically using the basic laws of quantum mechanics. Dyson and Lenard began with the mathematical formulation of the Pauli principle: the Schrödinger function had to be completely antisymmetrical in all N coordinates and this is where they got entangled in the most difficult geometrical problems. We were confident that our physics intuition would point us in the right direction.

Without the Pauli principle, the energy would increase greater than with N. First it was necessary to understand what the Pauli principle prevented to be able to recognize the danger. If the median distance between two particles is r, then the Coulomb energy between them is ~1/r as long as the impact of the charges farther away average out in shielding. In this way, the potential energy of all N particles together is around N/r.

The kinetic energy comes about, according to Heisenberg, because every particle that is trapped in an area the size of R receives an impulse ~1/r. If n particles in the area maintain a distance of r from one another on average, then it must have a volume $N r^3 = R^3$, *or* $R = r N^{1/3}$. *Without Pauli, the particles get along very well with each other, and it's possible to leave them alone in the community center without any further restrictions. When there is r, the impulse distribution will become smaller and smaller as N grows,* $1/R \sim N^{-2/3}/r$. *For all N particles together this is now* $N^{1/3}/r^2$. *Fig. VII shows the area of minimal energy in the phase space when* $R = 4$ *and therefore the impulse is ¼. This situation with, let's say, four bosons is comparable to four fermions. The Pauli principle applies to them; they are the ones insisting on a private room. It will then be R still having only a fourth of its previous value, and the impulse is four times as great. This means the minimal energy for bosons* $4 \times (1/4)^2 = ¼$, *for fermions* $4 \times (1 \times 1)^2 = 4$.

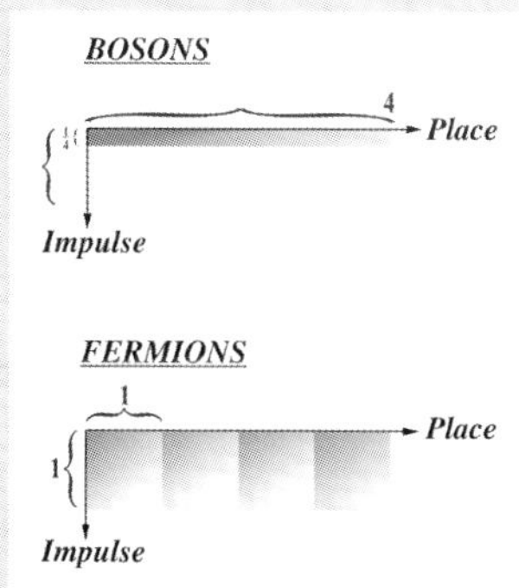

Fig. VII: Particle distribution in the phase space

The problem now is finding the minimal energy when we add the (negative) electrostatic energy $\sim N/r$ *to the (positive) kinetic energy* N/R^2. *N is a given and r can be adjusted, so that the minimum is accepted. This can be N-dependent, and we were ultimately interested in the resulting N-dependence of energy. For bosons, the kinetic energy was* $N^{1/3}/r^2$, *and is only able to balance the electrostatic energy* N/r *when r shrinks with N,* $r \sim N^{-2/3}$, *but then both energies are like* $N^{5/3}$. *This would mean that two gallons of gas would only* *fill up a quarter gallon, but would not have only twice as much, but something like 3.3 times more energy. This mega bonus is no longer valid under the Pauli principle, the electrons insist on their private rooms $r = R$. For minimal energy, the kinetic as well as the potential energy are $\sim N$, and that is what we need for stability. The story is easily told, but the challenge is pouring it into exact mathematical relations.*

Fig. 18: Elliott Lieb (on the right) and myself at work

Years passed without much progress. I had returned to Vienna in the meantime and had been honing my skills on related problems, but stability remained just over my head. In 1974, I could finally appoint a visiting professor, and I invited Elliott to Vienna. He was pleased to accept the offer and his time here was very stimulating, but the stability of matter resisted all of our attacks. We worked on it for another year, and in 1975 Elliott came to Vienna again. When I picked him up at the airport he said, "This time we are going to lick the problem, I can feel it in my bones."

Weeks flew by and we filled board upon board with countless formulas which were then all erased for being useless. Shortly before Elliott was to return home, we hit the bull's eye and there it stood before us on the board: what was to go down as the Lieb–Thirring inequality in the literature. An inequality which says that when the Pauli principle is applicable, the kinetic energy will always be greater than what is being said in the argument above with the private rooms. With its help, we were able to prove that the total energy is limited by the energy in the Thomas–Fermi theory. This theory is a classical limit and Elliott had previously proven, together with Barry Simon, that within it, as in classical mechanics, matter crumbles. This means that the stored energy can never be greater than what has been stored in the individual atoms. This energy is, however, finite in the Thomas-Fermi theory and the actual energy must be within it. The quantum mechanical binding energy among atoms can never be more than how much the Thomas Fermi theory overestimates the energy for the individual atoms.

After this success, I got caught up a bit in wishful thinking. Our proof had shortened Dyson and Lenard's from 40 pages to three, but we had used an inequality at one point that wasn't completely exact. We left it to cleverer people to prove the exact form of the inequality so it is now considered the Lieb-Thirring conjecture. Although it is already over 40 years old, and has been assailed by legions of mathematicians, it has held fast against all attempts of either verification or falsification. I therefore allowed myself to add it as an honorary axiom to the other axioms of the analysis. This means that the chemical binding energy of a conglomerate of atoms is always smaller than the sum of mistakes inherent in the calculations made using the Thomas–Fermi theory about the energy of the individual atoms. This is a more precise version of the statement that matter is stable in this sense.

Working together with Elliott went so well because we both had the same goal before our eyes. Our fundamental law was the Schrödinger equation and we only accepted what could be derived from it using mathematically legal means. Otherwise, we let our imaginations run wild; we explored everything from this perspective, from the stability of the stars to the chemical forces between the atoms. The latter is perhaps not so spectacular and comprises the content of our least quoted article. However, I'd like to talk about it a little, as there are some interesting and amusing things about it.

Right at the beginning I'd like to clear up a misunderstanding: there are no chemical forces. In "reality" there are only electrostatic forces which the atomic nuclei (positive charge) use to attract electrons (negative charge), while the electrons, as well as the nuclei amongst themselves, use them to repulse. The thing is that these charges position themselves in such a way that the attraction prevails. Normal chemical bonds are created when the atoms penetrate each other and the electrons primarily bandy about among the nuclei, so that they rather tie them up. Other than this kind of force, there are the so-called van der Waal forces even when the atoms do not penetrate each other. They are tailor made for the propagation of life. For the copying mechanism in DNA you need forces that:

(1) remain effective over great distances so that the bases can find each other,
(2) always attract, so that there are no gaps
(3) are not too strong, so that after copying these bonds can be undone like opening a snap.

The van der Waal forces can do all of this thanks to the gift of entanglement from quantum theory. This deals with the correlations that cannot be explained in classical mechanics and only exist in quantum mechanics. In the related article from Lieb and me, it follows that atoms attract one another by entangling. If I can't explain it, I'd at least like to describe what happens:

Let's start small with two hydrogen atoms. If we pull both of their charges, such that the proton (positive charge) and the electron (negative charge) are apart from each other at a distance d and put them on two lines with the distance a, then it's best to do so diagonally. Then the opposite charges come closer to each other than those with the same charges:

$$a\left\{\begin{array}{l} \overbrace{+\ldots-}^{d} \\ -\ldots.+ \end{array}\right.$$

The distance between the opposite charges above and below is, according to this, a, while the distance between the same charges is according to Pythagoras $(a^2 + d^2)^{1/2}$, which is greater. This is how attraction prevails. But is this the only possibility? Certainly not. The configuration above and below can be switched around, I can even turn it a little, but the argument remains the same. This condition is crucial in quantum mechanics because it tells us that electrons don't like being told where to go. They become petulant, begin thrashing about wildly and this costs a lot of energy. It is the best thing for the energy to specify as little as possible. We just need the charges to be diagonal to each other, nothing else matters. In classical physics, it's always necessary to show your true colors and say which particle is to be found where. In quantum mechanics, we can entangle the atoms; we just have to specify the correlations between the particles. I can't explain why this is, only Feynman knew. He said, "Nobody understands quantum mechanics."

Elliott Lieb and I were able to use an entanglement to show that neutral atoms attract each other over large distances in every case. The fundamental thing in our work here was its general character, because then the following question arose. It was now possible to produce anti-matter in which all charges carried the opposite sign. Two anti-atoms attract each other, because when I switch around the signs of all the charges it doesn't change anything about the forces. But what is the case between an atom and an anti-atom?

When in electrostatic force I change the sign of only one charge, then attraction becomes repulsion, and vice versa. If the van der Waals forces were to be repulsive, then we could store anti-matter in a container, because it wouldn't go near the edges to destroy itself. It might even be possible to collect enough anti-matter to produce an anti-matter bomb with unimaginable explosive power. If someone were to destroy anti-matter with matter, the reaction would be 100 times as powerful as in an atomic bomb. The explosive device would not weigh around 10 kg, but only about 100 g. No wonder that the US military were interested and invested money in this kind of research. There was even one researcher who, perhaps to satisfy his sponsors, fooled around long enough to get a repulsive result. Elliott and I said to ourselves that something had to be wrong with it. We started looking for proof that attraction would always prevail. We were successful even without a lot of money. Not only did we not have money from the military, we also didn't need modern analysis. The basic notion our proof is founded on comes from Newton!

P.S. The following is Elliott Lieb's version of events (from ESI-News April 2007)

Doing Physics with Walter
By Elliott Lieb

My scientific life has natural dividing lines, like new chapters in a book, the most important of which is the day I started to work with Walter. Walter says we met in 1968, and that is undoubtedly true, but my memory goes back to a turning point in the early seventies when we were sitting around a lunch counter somewhere and Walter asked me if I ever thought about the Dyson–Lenard proof of the stability of matter. No, I hadn't really, but he had, and he realized that, while correct, it needed some new mathematical insight to make it physically understandable as well as mathematically correct.

Walter invited me to be a 'Schrödinger guest professor' at the University of Vienna in the summer of 1974, and while this visit led to lots of fruitful scientific discussion (including collaboration with Heide Narnhofer) nothing dramatic happened except that I lost a key to the Institute. Fortunately, and this must be recorded for posterity, the mainstay of Walter's group, apart, naturally, from his wife Helga, was his assistant Franziska (Franzi) Wagner. She could do everything; not only type up the stuff we generated but also figure out how to deal with a missing, priceless, irreplaceable key that was official government property and must never, under any conditions be lost or duplicated.

Fig. 19: Franzi Wagner

But to return to the story, I was a 'Schrödinger guest professor' again in the summer of 1975 and this time we solved the problem. The main new idea was the realization that the kinetic energy of electrons (and other particles satisfying Pauli's exclusion principle) is always greater than the integral over all space of the 5/3 power of the particle density. This, in turn, meant that the old theory of E. Fermi and L.H. Thomas gave a lower bound to the total energy of matter, and this bound was known (by earlier work with Simon) to satisfy the desired stability condition.

We had several collaborations after that, but our last published work together was the proof in 1986 that the attractive van der Waals force between pairs of atoms or molecules, which drops off like the negative sixth power of the distance between them (in the absence of electromagnetic propagation corrections), was a universal consequence of Schrödinger's equation.

There were enjoyable visits before and after 1986. I would visit Walter in his hideaway in Zweiersdorf, where Helga was a superb hostess, and he would often visit my wife and me in Princeton. Walter is one of the most organized people I know. The visits followed a rigorous time schedule starting around 6 am and ending around 9 pm that dictated when we would wake up, take walks, play music, exercise and do science. Walter is an accomplished piano and organ player, and composer, while I was a fifth rate recorder player. Following an internal alarm clock Walter would announce that it was time to stop the science and play some baroque sonatas, and he would very patiently overlook my missed entrances and other mistakes. It was fun and I learned a lot, but I also realized that I would never be organized enough to do all the many things, in many fields, that Walter is capable of doing.

Unforgettable Characters

Scientists really are a motley crew. Everyone has their own working style and their own strengths and weaknesses. That's why people like to pigeonhole them and grade them — A is better than B — making some into heroes while others are vilified. I think that this is rather missing the point, everyone contributes in their own way to constructing the house of science as a whole and each kind of progress should be welcome. I have met all kinds of people over the course of my many years in research. Now, I would like to pick out three characters at random and relate what became of them. They were completely different sorts, and using their example, I would like to show how Fortune carelessly distributes accolades and obscurity, success and failure.

Arnold Sommerfeld

Fig. 20: Arnold Sommerfeld

Sommerfeld was the undisputed master of theoretical physics in Germany in the period between the wars. He created a school in Munich which produced leading figures such as Heisenberg, Pauli, Bethe, and Heitler among many others. He was the major influence on the development of physics and helped Germany become dominant in this field up till 1933. This did not prevent the Nazis in power from prohibiting him from entering his former Institute. Nonetheless, I never heard him utter a bitter word about it. He knew what he was worth and stood far above haggling for public recognition. I consider it one of the strokes of luck I've had in my life that I had the privilege to get to know him personally. He was already over 80 at the time. He had a small, wiry stature; his mustache was that of a Prussian officer and was perhaps revealing his Eastern Prussian background. His success lay in his charisma, the openness with which he treated people.

I was still a student at the time, but I found myself immediately involved in scientific discussions with him. He took it for granted that I knew something about the topics current at the time. In particular, Herbert Fröhlich, the solid-state physicist who was quite well-known then, had proposed a new theory of superconductivity which Sommerfeld was fascinated by. He was one of the founding fathers of solid-state physics and finally here was a proposal which functioned within the known laws of physics. This phenomenon was so peculiar that it was often supposed to be invoking new laws. For example, it didn't seem to make sense that a current could flow at zero degrees where there is no direction given at first. This, in turn, led to the creation of the term of spontaneous symmetry breaking. Sommerfeld apparently had the feeling that if there weren't any new laws, then at least there should be new terms on the way.

His interests were not limited to physics. He once remarked of existentialist philosophy, which was very popular at the time, "That isn't even philosophy, it's just a declaration of bankruptcy of philosophy."

Sommerfeld's scientific achievement has always been poorly classified up to now, as it is located too close to the boundary between physics and mathematics. He became famous for the fine-structure constant which bears his name. This measures the strength of the electromagnetic interaction without relating to a certain system of measurement. It's just that coming up with a constant is not quite a scientific achievement, and God knows that Sommerfeld had more to offer; for instance, the formula for the fine-structure of the atomic spectral lines. This was at first a success, then an ignominy, then a huge success once more. Sommerfeld had calculated the influence of the special theory of relativity on the atomic spectra within the framework of the old Bohr theory before there even was a real quantum mechanics. He found a formula for this which best represented the empirical material.

This was considered to be the first confirmation of Einstein's theory, because at that time high speeds were produced by letting an electron plop down on an atomic nucleus. However, it was later discovered that electrons have completely different properties than those that Sommerfeld used in his calculations. In reality, electrons have a spin, which is not mentioned in Sommerfeld's work, and they conform to a wave equation, not the Newtonian motion equation.

Schrödinger was the first to study the wave equation which satisfied the requirements of the theory of relativity. It is now called the Klein–Gordon equation. If you solve this equation for particles without spins in the electrical field of an atomic nucleus as Sommerfeld did, then you come up with a different, and therefore empirically disproven, fine-structure formula. That is why Schrödinger didn't publish these results, but just the limit of small speeds, the Schrödinger equation. Much later it was possible to build particles without spin and then the false fine-structure formula was right. Dirac had the brilliant idea to take the root from the operator for the Klein–Gordon equation and, miracle of miracles, the result fit a particle with spin 1/2 (Dirac equation). If you were to solve the equation for a particle with spin in the electric field of an atomic nucleus, then you'll come up exactly with Sommerfeld's fine-structure formula. So the wrong dynamic for the wrong particle delivers the same formula as the right dynamic for the right particle.

Sommerfeld was not only fate's plaything of ironic coincidences; for example, his diffraction theory was most extraordinary. The heart of the matter is mathematical finesse, and that is too heavy for most people to stomach. The problem that he solved is the diffraction of a wave on a wedge. There had been a "theory" about this before he came along, but it was based on improvable suppositions and no one could say how great the miscalculation made was. The result was the small difference between many positive and negative charges. The answer varied depending on which were taken into account. Now, Sommerfeld was able to solve the problem with precision and it was revealed to what extent the previous theory had been in error. Sommerfeld's method of exacting mathematics became the lodestar guiding my professional activities. A volume of Sommerfeld's lectures which he personally signed with a dedication to me is therefore one of my most treasured possessions.

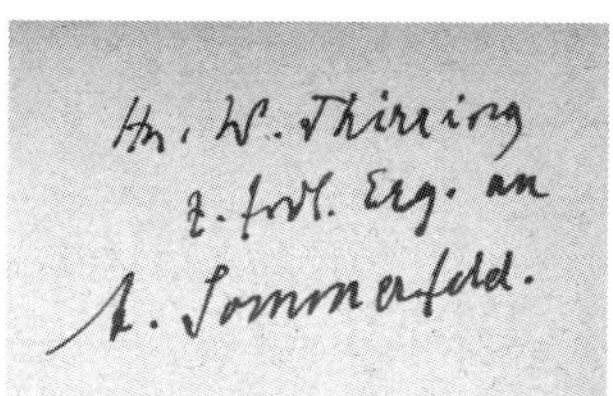

Fig. 21: Dedication from Sommerfeld

Guido Beck

Fig. 22: Guido Beck

It was a beautiful spring day, a bit of green dared to peek through, but the sun was still listless. A small group of mourners had gathered near an open grave into which a modest urn was being lowered at the cemetery in the town of Kasten bei Boeheimkirchen. During the solemn requiem in the church, all of my memories and emotions came flooding back. I met Guido Beck for the first time in 1960 when he suddenly stood before me in my room after I had given a lecture. He was an older man of small stature; his skin was leathery from his constant smoking, but he had an irresistible charisma. He wanted to know everything about the state of science in Austria, and he invited me to come to South America. After just a few minutes of talking, we were already talking shop. It took a long time for me to be able to reconstruct his biography and recognize him for who he was.

What I found was this: Guido Beck was born in Reichenberg (now Liberec, in the Czech Republic) in 1903. He attended secondary school in Switzerland, and after graduating he studied physics in Vienna with my father. This is where his talent began to unfold, and he quickly mastered the general theory of relativity. He was the first one to be able to offer an exact solution to the Einstein equation which represented gravitational waves. However, this theory wasn't popular at the time, and so no one really paid any attention to his work. Still, at least Heisenberg was so impressed with him that he made him his assistant.

Of course, Don Guido (this name became established over time) had to change his area of study, and since then he dealt with quantum theory, in particular with nuclear physics.

After 1933, being a Jew he was forced to emigrate, and his odyssey began. His reputation had reached all the way to Kansas and he was offered a professorship for two years. However, Don Guido was a highly sophisticated man; for example, out of respect for the locals he always learned their language, and the American Midwest did not exactly fulfill his cultural requirements.

So, he returned to Europe and wherever he came, students flocked around him and he advised them scientifically and personally. His admirers and their descendants can be found scattered all over the world. You can find them in Odessa, which he was able to flee from Stalin's terror just in time, in Lyon, where he supported Jewish refugees from Eastern Europe, he was in Copenhagen, in Prague; he embodied the unified European. Then the war began limiting his sphere of action more and more. He finally wound up in Portugal, and realized that South America was his destiny. Europe had become too restrictive, and he loved to travel. Once when he had been invited to give a lecture in Ukraine, he was told with regret that the travel expenses could be reimbursed only within the borders of the Soviet Union. Don Guido was able to use this to his advantage by arriving in Vladivostok and, with the help of the Trans-Siberian railway, he was able to form his own impression about the size of the country.

Therefore, when he received an invitation to the observatory in Córdoba, Argentina, in 1943, he stopped teaching in Portugal and set sail. He arrived at the harbor from Buenos Aires with one dollar in his pocket. He had absolutely nothing superficial or materialistic about him, and he projected such energy and abandon that he had soon conquered the academic world in Argentina.

He remained in Argentina for eight whole years, which for him was a very long time. In the meantime, news had spread about his scientific work all over South America, he had naturally added Spanish and Portuguese to his treasure chest of languages, and in 1951, he accepted a professorship in Rio de Janeiro. By the time I met him, all of South America had become his home, and he spent some time here and some there.

Until today, Don Guido is a kind of legendary figure, and there are many anecdotes circulating about him. Not only is his professional, but also his personal life worthy of mention, as it was not without heart wrenching aspects. On a visit to Copenhagen, he had fallen in love with a Danish woman and they had gotten married. However, in the chaos of war they had become separated. He went to South America and could not return to Europe for a long time. When this was finally possible, he wanted to tie up any loose ends and officially divorce the wife from whom he'd been separated for so long. So he came to Europe and met with her. As they went through the divorce papers and came to the point of grounds for divorce, they realized that they didn't have any. He took her with him back to South America. However, their happiness was not to last, because she soon died of cancer.

Don Guido introduced me to the beauty of South America, which truly enriched my life. I once visited him in Bariloche, where the Argentinean government had established a research center on a beautiful lake in the Andes. The accommodations are quite rustic in block houses, the restaurants are simple, yet refined, which could be seen; for example, in the fact that in one Bach's Brandenburg concertos could be heard in the background. Another sign of the fine culture was that, although Don Guido had no car, he had a horse. I was also able to borrow one, and we explored the area on horseback.

I observed another manifestation of South American temperament once when I visited Don Guido in Rio. It turned out that Mathilde von Pflügel, a great-granddaughter of Christian Doppler, was living there, and her birthday was being celebrated in a restaurant in the middle of the jungle. After the usual food and drink, they wanted to give everything an Austrian touch, and so they all danced a Viennese waltz. The guests danced correctly, but it was obvious that this dance did not correspond to their mentality. After Mathilde and Don Guido had vacated the dance floor, someone put on a samba, and it was like lighting a fuse. Not only the posture, but also the expressions were completely different. It was like they had suddenly fallen into some kind of intoxicated state and it was impossible to ignore this ecstatic drive.

The reader might be asking, but why are Don Guido's ashes lying in Boeheimkirchen, of all places? An acquaintance of his had an estate there and, after he was killed in a tragic car accident, she along with Mathilde von Pflügel organized the transport of his urn. So, at least he found his last resting place in Austria.

Felix Ehrenhaft (1879–1952)

Fig. 23: Felix Ehrenhaft

During the Nazi period racial prejudices were being rammed down our throats and all sorts of things were being attributed to the Jews. For example, it was said that Jews weren't athletic and the counterexamples like Paul Preuss, the pioneer of free climbing or Bob Kronfeld, the pioneer of gliders along with many others, were already being forgotten. The only other competing theory at the time was the resentment Austrians had toward Italians. As these things happen, my first scientific assistant, Sergio Fubini, was both, namely Italian and Jewish. Not only was the characterization we had learned completely wrong, he even had the audacity to be blonder than I was. To top it off, he disproved the so-called Jewish egomania; you couldn't call him anything other than an unwavering idealist. He devoted his remaining energies to organizing joint seminars for Jews and Palestinians. Unfortunately his efforts, like those of Daniel Barenboim, have had no political effect.

I have gotten to know many Jewish people in my life, and in many ways Felix Ehrenhaft best personified the realization of the abstract stereotype of a Jew. He often aroused animosity, but I don't see him in a completely negative light. I would rather say that he did more for Austria than Austria ever did for him, and I would like to take this opportunity to honor him. He was a physicist, and although abstract mathematics wasn't his forte, he had a keen sense of what was really significant. Already as a young researcher he came up with the idea that would have ensured him a seat in the front row of physics if he had developed it with sufficient precision.

Ehrenhaft discovered the standard method of measuring the electric elementary charge. In doing so the charge is put on a suspended particle and it can be seen how the movement of the particle can be influenced by an electric field. Today, this is called Millikan's oil drop experiment, because Millikan used oil drops for his suspended particles. These have the advantage of being completely round, and there's a nice formula determining their drag due to the air.

For his experiments, Ehrenhaft was using metal shavings, but these come in such bizarre shapes that no one knows what their drag is. So he lost the credit for this experiment, already considered a classic, even though he was first.

The thing was that Ehrenhaft was a sore loser, and instead of admitting his mistake, he reacted with arrogance. With his behavior he discredited himself internationally, and when he was forced to emigrate in 1938, he was ignominiously barred from entering the academic gates in the USA. In the meantime, he had learned how to convince financially powerful circles of his importance and was in this way able to keep his head above water. In any case, he returned to Austria in 1947. His Institute, which he had directed until 1938, had been liquidated by the Nazis, so that he worked as a visiting professor from America. At first glance this sounds quite good, but it turned out to be fatal.

He acted like a television host in his lectures. For example, he liked to begin his presentation with the line, “I and Faraday have discovered…” He was quite concerned with his appearance at public lectures and would ask his secretary before going on, “Should I let my hair stand on its end all wild like a genius, or slick it back like a gentleman?” This brought the poor lady in a difficult situation, as he was bald. She was able to find a diplomatic solution and told him, “Professor, we all know that you are a genius, look like a gentleman.”

Ehrenhaft was always concerned with outward appearances and even in 1913, when Kaiser Franz Joseph opened the new Institute building, he knew how to maneuver himself to the front row. My father, even though he was his scientific superior, wound up in the second row. This was for the best, as the old monarch wasn’t interested in physics. He simply asked Ehrenhaft, “Did you serve?” upon which Ehrenhaft stood at attention and exultantly exclaimed, “Yes, Your Majesty, regiment …” Immediately, His Majesty’s full attention was directed to the professor’s concerns.

Over time, Ehrenhaft perfected his fundraising ability to such an extent that he was able to lure Einstein to give a lecture in Vienna for an incredible fee. It was rumored to be a five-figure amount in schillings, which would be about 50 times the monthly salary of a professor. He defended the amount to me with the argument that, “Einstein surely has to get as much as a tennis-boxer.” He had projected all top athletes into an abstract space and made his calculations accordingly.

Today there’s a plaque on his former villa in the Grinzingerstr. 70 which still commemorates this event. Here is where Ehrenhaft held court in great style; for example, he would have the chef from Hotel Sacher come to prepare dinner. People started asking how a university professor could afford to live on such a grand scale.

Fig. 24: The menu from a colloquium banquet

Maybe it's true what he once told me, how the general director of a major concern paled as he heard Ehrenhaft's expert fee demands after a patent trial.

Ehrenhaft's guest book is a Who's Who of the upper class in science. His villa was not only a meeting place for Nobel Prize winners, but also a gathering spot for representatives of the fine arts. Bertha Zuckerkandl, whose salon welcomed all of Vienna's intellectuals, can be found in the guest book as well as Alma Mahler-Werfel, whose favors were evenly distributed amongst the different art forms. It would go beyond the scope of this book, and distort its balance, to list all of the names here. However, I can't refrain from reproducing an entry by Einstein, because it's such a good expression of his waggish humor. After a typically opulent meal which followed his lecture,

Einstein was able to hammer out the following verse about the social whirl:

Denkt auch manchmal an den Alten	Think about the old peer
Der einst Predigt hier gehalten	Who used to preach here
Drauf Frau Hofrat unerschüttert	And then the unshakeable court counsellor's wife
Hat die Bonzenschar gefüttert.	Fed all his cronies the best meal of their life
Mann, Weib, Tochter und der Sohn	Man, wife, daughter and son
Liefen stets zum Telephon.	Were constantly running to the phone
Frassen selbst bei Diplomaten	Even ate at diplomats'
Alle restlos freundlich thaten.	Everyone friendly, tipping their hats
Hörten Piccard auch berichten	Listened to Piccard also tell
Seine grusligen Geschichten.	His horror stories full of woe
Schönes hab' ich viel erfahren	I found out lots of nice news
Werd's im Sinne stets bewahren.	And will always have them to use

It can't be denied that Ehrenhaft had a certain talent for recognizing people's abilities and interesting problems. The students who worked with him such as Peter Weinzierl, Michael Higatsberger, Franz Viehböck (the father of the cosmonaut) and myself later became the bearers of physics in Austria. The problem that he concentrated on was naturally out of his league, but continues to fascinate people: the discovery of magnetic monopoles. He didn't care that these were prohibited by the theory of the day. According to what we now know, they are a theoretical possibility, but haven't been discovered yet. In any case, Ehrenhaft knew how to turn his work on this problem into a production. His beginning assumption was correct; if there are magnetic monopoles, then they can only be produced by cosmic radiation. If you want to find them, you have to reach for the stars. He therefore had himself carried up the Sonnblick mountain in the Austrian Alps in a sedan chair. Of course, he couldn't do any real science up there, but it had a great media effect.

Ehrenhaft always behaved generously towards us; he took me to Como with him to the Volta conference in 1947, for example. That's why I was so outraged when I discovered in the university archives how the Austrian government dispensed so little courtesy toward him. In 1949 he turned 70, and he applied for retirement benefits stating that he had been working at the university since 1903, and that his dismissal in 1938 had been illegal. His application was rejected, and the reason given was that retirement benefits were not intended for visiting professors. He had to work for two more years, up until he died.

Contemporary Mathematics in Physics

Soon after I came to Vienna in 1959, the mathematician Leopold Schmetterer also returned to the University of Vienna. Schmetterer was a widely educated person and immediately signaled his interest in our work. We reciprocated his interest and decided to hold a joint seminar. The only purpose of the seminar was the exchange of information, and first each had to learn the language of the other. Therefore everything was very informal, and no question was too dumb to be posed. This is how I slowly began to understand what was going on in modern mathematics and it dawned on me that my math and I belonged to the wrong century. The problems in modern physics can be much more precisely expressed in the language of 20th century mathematics than in that of the 19th.

My younger colleagues had also come to this realization and paved the way for me. Roman Sexl recognized that the mathematical language of Elie Cartan was perfect for Einstein's general theory of relativity, and Alfred Wehrl configured quantum mechanics using the theory of von Neumann's algebras. I revised my lectures accordingly, after all my audience were going to become the professors of the 21st century, and I couldn't have them speaking the language of the 19th century. Finally, I expanded my lectures and had them published in book form as a four-volume course on mathematical physics. Naturally, their reception was somewhat tepid at first, because it meant learning lots of new things. But I also received international recognition for them. When an international association for mathematical physics was founded, I was elected as its first president in a worldwide election. Meanwhile, this language has become the general standard.

After Paolo Budinich founded his national school SISSA in Trieste, my colleagues regularly went there for years to lecture on my tome on mathematical physics. It still fills me with pride when I see what extraordinary scientists have developed from this. Unfortunately, my colleagues mentioned above did not live long enough to see much of the fruits of their labor.

I will here try to give the reader an idea of what kinds of thoughts were being spun at that time. Because this again requires a bit more consideration which I do not wish to burden every reader with, I have put it all in the box below. The point I am trying to make is showing that abstract language does not automatically whisk away thought processes to an esoteric realm, but rather generalizes them so that they can be found everywhere in life.

The intellectual tools used to cope with the infinite were hammered out by the genius of Georg Cantor. Unrecognized in his lifetime, today every mathematician speaks his language, which as set theory has become a cornerstone of modern math. It is the language of pure logic and is what Nicholas Bourbaki spoke in his undertaking of a purely conceptual construct of mathematics in the 20th century. This produced a work of 32 volumes (as many as there are Beethoven Sonatas) he called Elements of Mathematics (like Euclid), in which he standardizes mathematics into a network of logical structures. Bourbaki's worldwide success achieved the same level as that of Euclid; if it will also last for 2,000 years remains to be seen.

We don't know much about the person Euclid, but about the person Bourbaki we know for sure that he never existed. Behind this pseudonym hides a group of brilliant French mathematicians who didn't want to accept the fact that international mathematics would wander over to Germany to Hilbert. They wanted to lay the foundation themselves and have done so in a public seminar developed over decades. However, it is simply too much to demand of a non-mathematician to plow through 32 volumes, and, in order to reach a wider public, one of the "Bourbakists", Jean Dieudonne, composed a nine volume extract.

I was already taken with the first volume; it was exactly what I needed for my mathematical physics, which is why I contacted Dieudonne and asked for the next installment. However, Dieudonne was still correcting it, so I offered to help him with the proofreading. Dieudonne accepted my offer, as he realized that I was a perfect guinea pig being an interested, but largely ignorant reader. I dutifully carried out all of the calculations, and we struggled with signs, 2π factors and other, similarly bothersome stuff. I knew that I had two tasks. On the one hand, I needed to play dumb when something wasn't sufficiently explained, and furthermore I had to act according to the principle that whatever Dieudonne has written is wrong, and I must construct counterexamples. There was ample opportunity for the first, because when someone has worked with a language for years, they lose the feeling for what an outsider can understand or not. For example, at one point he sprinkled the words "short exact sequence" in and assumed that this was something everyone had heard starting in the cradle. I exclaimed that I hadn't, and forced him to say that that was a series of transformations in which each transformation had the image of the previous depicted in a distinguished element. I could accept this definition, we had already come across it. My second function was also not without its usefulness: I was able to discover a real mistake. I will describe this in further detail after I have first provided the reader with the necessary preparation.

I understood 19th century mathematics to be what we had learned in school, things like simple functions. You say that the coordinate y is a function of the coordinate x, both ordinary numbers, when you specify how to calculate y from x. This is usually written as y(x). A simple example of this is $y(x) = x^2$

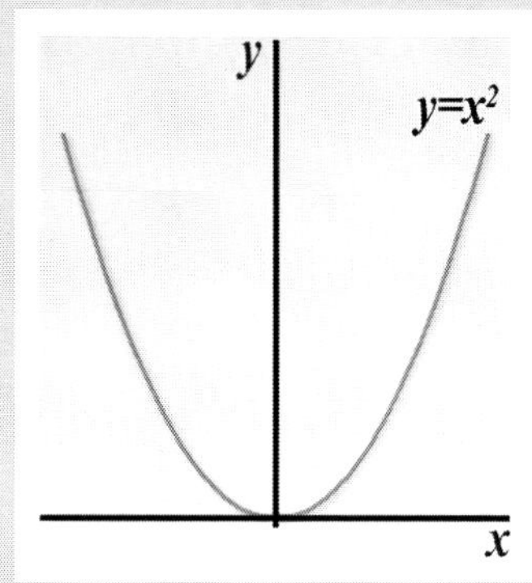

Fig. VIII: The parabola

When you regard x and y as coordinates on one plane, then you can use the points in the equation $y = x^2$ *to draw a nice parabola. I call it a nice one, because this curve has the property of lying over each of its tangents (this is called strictly convex).*

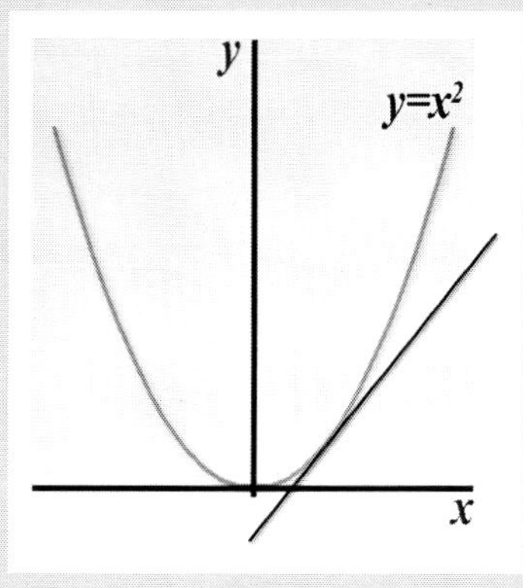

Fig. IX: An additional tangent

It follows that it can have at most one minimum, because if it were to have two and you were lying on one of the tangents, then everything else would have to lie above it, including the second presumed minimum, and then it couldn't be a minimum any more. In our example the minimum is actually at $x = 0$ and has the value of 0. There are no maximum values in sight, they would be at $x = \infty$ or $x = -\infty$. But I could allow x to vary only between 1 and -1, and then there would be two maximums at these final points. In any case, they would have to lie on the edge, because if the parabola were to continue after the maximum, then higher values would also have to exist beyond it. The attentive reader will have noticed that the fact that x is an ordinary number was not mentioned in the argument. x could also be on a plane for both coordinates, y(x) would then be a mountain. Indeed, x can be anything for which the tangential plane can be defined. With this we have proven the following general statement: ***the minimum of a strictly convex function is unique. There can be any number of maximums and they are all located at the edge.***

Modern mathematics takes it a step further and allows y to also have some element of a "set". The term function is generalized to a transformation of the set X onto the set Y, meaning that each x out of X is assigned a y out of Y. It can happen that two different xs are assigned the same y. In our example, +x and –x had the same y. If this doesn't happen, the transformation is called ***injective****.*

Even when each x has a corresponding y, it doesn't mean that there has to be an x for each y, so that $y = y(x)$. For instance, in our example all ys are positive. If all ys out of Y are present, then the transformation is called ***surjective****, and* ***bijective*** *when it is both injective and surjective. The words image and preimage are easier to understand. The former are the elements which appear as the result of the transformation, and the preimage of an element are the elements that it delivers as an image. So, surjective means that the image is the complete set, and injective demands that the preimage of each element consists of only one element. Using these terms we can capture the abstract characteristics of a transformation, they do not relate to what is transformed where.*

We have now gotten three subsets of the sets of transformations; the injections, the surjections and the intersection of these, the bijections. Roughly speaking, Y must be greater than X to have an injection and X has to be greater than Y to have a surjection.

X and Y can also be the same set, then there will certainly be a bijection, namely the transformation that assigns each element to itself. Conversely, a bijection of X to Y is an identification of their elements and X and Y can be seen as being the same for many purposes. In any case, the inverse x(y) is also a transformation of Y to X.

This is our segue to another important mathematical term which has also been liberated from any earthly imperfections, namely the term "group". All elements of a group can be multiplied and every pair is assigned a further element. Of the products of numbers, we only take over that ((a times b) and all of that times c) results in the same as (a times (b times c)), so that we no longer need the parentheses and can simply write abc. Otherwise, the definition is quite generous: bijections form a group and they can be executed successively.

This all sounds terribly abstract; therefore, I would like to show how this is really about normal, unsophisticated everyday ideas.

The simplest way to illustrate these terms is if the sets X and Y only contain a finite number of elements, like all of the integer numbers of an interval. We can imagine that X are men and Y women and transformations are marriages. Every man gets a wife. Bijection means that everything's moving forward; there are the same number of men and women.

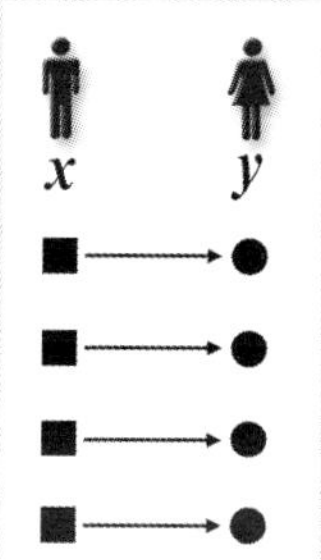

Fig. X: Bijections as marriages

In this case, injective and surjective are equivalent; if every woman has just one husband, then it will end equally, and every woman will get a husband. If things are reshuffled and every woman still has just one husband, then every woman will still get one. The bijections are then exactly just the permutations. In general, we see that when there is a bijection between two finite sets, then the three terms injective, surjective and bijective coincide, despite their generality. If there is no bijection, then the transformation term is here "feminist": a woman can have more than one husband.

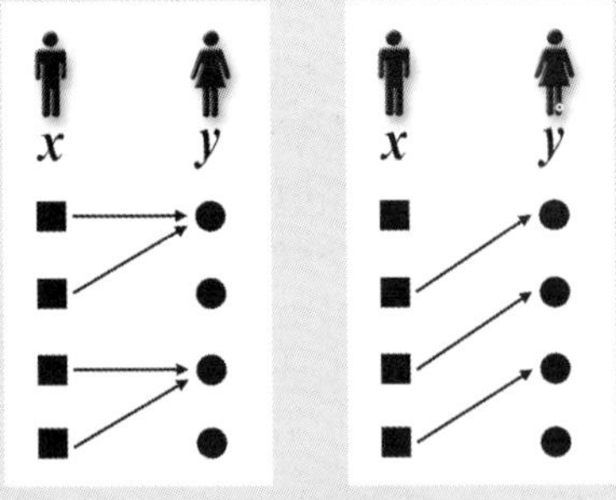

Fig. XI and XII: Marital difficulties

On the contrary, polygamy is prohibited.

Whoever was paying close attention will have noticed that this cannot hold for any random set; the function $y = x^3 - x$ *from the real numbers to the real numbers is; for example, surjective but not injective, it has the value 0 when* $x = 0, 1, or -1$*.*

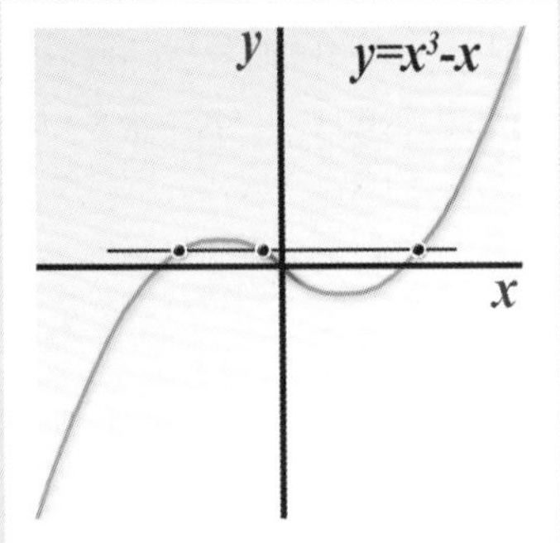

Fig. XIII: A non-injective surjection

Why don't our marital observations apply here? Is it because x and y are continuous variables? No, that's not it; it's their infinite stockpile of values that's making a mockery out of our reasoning. This can even happen when everything would be discrete. Hilbert illustrated this point by picturing an infinitely large hotel. The porter's task is also an injective transformation, each guest has to receive a room, but no room can be double booked. The hotel is full when there is a bijection between the guests and rooms. If anyone else shows up there is nothing to be done: the porter would have to find a transformation that is not surjective, because there should always be space for newcomers. It also has to be injective, but that can't be, not when the hotel has like most hotels a finite number of rooms. If the hotel were to have an infinite number of rooms, then one could easily find such a transformation. It just has to move the guest from room 1 to room 2, the guest from room 2 to room 3 and so on so that each guest still has their own room and room 1 remains vacant for the new guest. Even an infinite number of new guests wouldn't be a problem. The porter would just have to move the guest from room 1 to room 2, the guest from room 2 to room 4; to put it in general terms, from room n to room 2n. Then the odd-numbered rooms would remain vacant and ready for the new guests.

Infinity can really get to you. You must keep on your toes because it's always appearing under a different guise. We began with the simple tasks of a matchmaker or a night porter, and at the question whether the set B is as large as set A, we have met with unexpected paradoxes. Naively you'd think that A would have to have room in B, therefore, there must be an injective transformation from A to B. Injective because otherwise I could transform all elements of A into the same one from B, which would mean that B could consist of only one element.

A useful statement is certainly that B is the same size as A when there is a bijection from A to B. Their inverse is also a bijection, and just as we wanted, the statement that A is as large as B is also true. Furthermore, if C is as large as B, then it is also as large as A, because the sequence of bijections results in another bijection. It's better to abandon the hackneyed word "large" and to say that the two sets "are of the same cardinality" when there is a bijection between them. This does not in fact quite cover our understanding of size, because it can be that one part has the same cardinality as the whole. If each natural number is assigned its double, then that is a bijection between the natural numbers and the even numbers, although you could naively say that there are less of the latter. However, there are different kinds of infinity; infinite sets that cannot be bijectively transformed to one another; for example, the set of all real numbers has a higher cardinality that the set of the natural numbers.

These few remarks show how you should proceed. You should start from images that are clear without further explanation and consider them to be embryonic forms of more general terms. You need to stretch these as far as possible, so that they cover many situations, including some that are easy to follow intuitively and point you into the right direction of where there's more to discover. However, this universality comes at the cost of specific statements which could be possible under special circumstances. It is at your discretion as to how far into the abstract you want to go, and to know when you've just been expanding on pipe dreams.

Finally, I'd just like to relate how Hilbert's hotel helped me find a mistake while proofreading Dieudonne's book:

He had noticed in investigating the grounds of the transformations that the unitary ones were a strongly closed subgroup. I first had to translate that into something I could understand, but we already have mentioned the right catchword "group" above. Whatever it is that the unitarities transformed, they have to be invertible as group elements; in other words bijective transformations. Strongly closed means that appropriate limits of these elements also enjoy these properties. This situation presented me with the opportunity to show Dieudonne up, because the statement was so general that it should also hold for the physical processes I was familiar with. So you can; for example, take the current of water to be a group of transformations, with time as the group parameter. Surjectivity of the transformations then means that the water goes everywhere and injectivity is the incompressibility of the river. However, I can put an insurmountable obstacle in the river so that the water would flow back behind it and the area in front of it would dry out. Over a long period of time, it will get completely dry; the limit of large time spans in the transformations is therefore not surjective, as there is no longer any water there. This disproves Dieudonne's claim; the time limit introduces elements that don't belong to the group. They are what the porter had done in the endlessly large, but full hotel to keep a room vacant.

I immediately wrote to Dieudonne depicting these considerations upon which he ruefully admitted to his slip. He then went on to fix it in a callous way; he simply demanded more of the convergence which disqualified my counterexample and saved the claim that the set of unitary transformations is closed.

The Erwin Schrödinger Institute

The following reflections on our cultural heritage can be considered to be a display of historical consciousness or nostalgia; at least, these were my motives for my actions after the fall of the Iron Curtain and the changes that took place for our neighbors to the East.

The Danube monarchy, this melting pot of nations, produced an intellectually potent brew and intellectual life flourished during the last decades of its existence. Not only was Vienna traditionally a leader in music and medicine, but also in other branches of science, the fine arts, architecture, and city planning where immortal achievements were made. The paintings of Klimt still command top prices internationally, Freud's psychoanalysis is practiced all over the world and Boltzmann's formulation of the second law of thermodynamics dominates our way of thinking. Still, the most lasting achievements remain the three jewels for capitals: Budapest, Prague and Vienna, which are still competing amongst each other for the title of which is the most beautiful. They form a stylistic unit, but when I returned to Vienna, they still lay in different worlds. Somehow you got used to the Iron Curtain, but you could never really accept it. Many people had relatives across the border; in my case it's simply due to historical coincidence that I belong to the Austrian branch of the Thirrings. Why did they, just about an inch to the right on the map, have to live under so much worse conditions? The ideological fronts had become so deeply entrenched in their positions that it seemed impossible to reach an amicable agreement. I personally experienced this during a short political interlude.

When in the 1970's the Russians were building the SS-20 nuclear rockets, it was clear that this further increase in the nuclear arms race was a danger to humanity. So a small group of people from all walks of life; Wilhelm Dantine, a theologian; Thomas Schönfeld, a chemist; Alfons Stummer, a filmmaker; and I held a press conference to let the public know about this danger. Our goal was to have Austria propose at the United Nations that further nuclear arms build up should be banned around the world. In fact, there was a reaction in the media and politics, but it was across the board negative. Our initiative was dismissed as communist propaganda, because everything that was against nuclear weapons was considered communist. The fact that our initiative was primarily aimed against the Soviet Union was not considered to be very significant.

We were also received by Federal President Kirchschläger, who naturally did not repeat nonsensical slogans, but he thought that we were unworldly and delusional. The mistrust between East and West was so great that an agreement on nuclear weapons was unthinkable. But then Gorbachev came on the scene, and the ideological concrete began to crack. A treaty reducing the amount of nuclear weapons between the USA and the Soviet Union was signed. Hopefully not too late, because the reduction of 50,000 nuclear rockets to a few thousand is an enormous technological undertaking. This can be seen in the science budget in the USA, in which the large accelerators costing a few hundred million dollars are small fry; the costs for NASA reach in the billion range, but are still not the largest item. The largest chunk goes to maintenance of nuclear weapons. We can only pray that things are going well in Russia, which is financially weaker, because even a single rocket could cause immeasurable damage in the wrong hands.

Towards the end of the 1980's, small changes in the political blocs could be made out, but still the speed with which the fall of the Eastern bloc occurred was surprising. I experienced it firsthand:

Other than the Triangle Seminar, which took place three times a year in the three different countries, my wife and I enjoyed visiting Budapest and Prague for smaller conferences simply because we love these cities. Once we were at a meeting in Prague, and my restless legs began bothering me towards afternoon. The weather was gorgeous and I had dutifully listened to a few lectures, so I wanted to take a little walk around town. I carefully slipped out the back door. Now, Prague is a wonderful city to take a walk in. When you cross the Charles Bridge and continue through a bit of the Old City up to the Hradcany castle, you can enjoy ever changing views of the river and the Golden City. There are so many tiny, narrow streets curving along the way lined with old buildings, until finally the majestic castle with its courtyard resembling a city within a city stands before you. I savored the view on this afternoon as well and then found a small, exclusive restaurant where, for a pittance, I received the royal treatment. Although everything seemed so peaceful, there was a sense of foreboding in the air. Was that noise in the distance people shouting? People seemed to be upset somehow, and in any case, I thought it would behoove me to start making my way back. Upon arriving at the hotel, I was met with agitation and excitement. People were saying that a revolution was going on in town, and everyone had been worried about me. Thousands of people were demonstrating at Wenceslas square; the time was up for the communist regime. Most locals reported that at least one of their relatives had taken part in the demonstration. At first the turnaround was acknowledged by a kind of frightened joy, but after the news appeared to confirm that the change had taken place, an open sense of triumph mixed in with anxious hope.

When we returned to Vienna, the political map of Europe had changed. The mood was euphoric; the first step towards a unified, democratic Europe had been taken. Still, I had some sense of unease, and it seemed to me that it was imperative that we remain ready to help especially now. Up to now our colleagues had had it comparatively good; they came from a privileged class. This was particularly true for physicists after they had built Stalin an atomic bomb. These privileges would now come to an end, and also the general economic situation couldn't change from one day to the next. In the short term, things could only get worse for our colleagues. Naturally, they would be free and able to leave the country, but that was exactly what was to be avoided. At the end of the day, these countries had invested a lot in ensuring having a highly educated intellectual class at their disposal, and it would be extremely unfair if now other countries were to be the beneficiaries of this investment.

These were the ideas I was mulling over when one day Peter Michor, a colleague from the Mathematics Institute, came to me and reported that the Russian mathematician Vinogradov had written him to say that he wanted to advocate having an institute for non-linear problems in Vienna. Russian mathematicians who were now more mobile would like to meet and work together with their Western counterparts here. In order to give this project a chance of being realized, they wanted my support. I understood that I was needed now. Until then, I had always avoided academic posts with the excuse that others were better suited and would like to do it more, whereas for me the work-benefit ratio seemed much too high. But this project was worth it to me to lose a few years. First I wanted to redefine it, in order to give it more political sex appeal. Multidisciplinary was the buzzword of the time, so I dropped the word non-linear, after all, who wants just to solve non-linear problems — and replaced it with "mathematical physics". The aspect of cooperation across national borders had to be included, so we needed the adjective "international". Finally, if it should be located in Vienna, the main financing needed to come from Austria, so it needed to be named after a famous Austrian. Schrödinger had already gained a certain level of popularity through his appearance on the thousand shilling note, so I came up with the name the International Erwin Schrödinger Institute for Mathematical Physics. This name has stood the test of time, but whoever thinks it's too much of a mouthful simply says ESI.

Naturally, I realized that the founding of an Institute was more work that I could take on by myself, so I looked for comrades to share the load. Peter Michor was willing to help. I had to proceed with caution; however, not that anyone would have the feeling that there was any preferential treatment. So for the beginning it should be a physicist, preferably a woman. Luckily this woman existed: Heide Narnhofer. She had a good reputation internationally, and I was very pleased when she agreed. Heide Narnhofer, Peter Michor and I were a harmonious trio without petty jealousies; we didn't have time for that kind of thing. We got right down to planning.

Fig. 25: The founders (Peter Michor, Heide Narnhofer, and Walter Thirring) attending a lecture at the ESI

The following two years were a roller coaster ride filled with highs and lows, we would feel on top of the world, only to be shot down to the depths of despair. The feeling of anxiousness was quickly followed by relief, optimism alternated with despondency. However, it all comes down to success, so that I don't want to bother the reader with complaining about overcome obstacles, but rather pick out a few turning points along the way.

The first thing to do was procuring the necessary materials, and collecting data about similar existing institutions that we could use as our model. We calculated all of our costs and came up with an annual budget of 40 million shillings, which was a sum also considered adequate by the Ministry for Science. The next thing was to invite the leading experts in their areas to evaluate our project and then inform the Ministry of how they rate our chances for success. To our great delight, all the most important people were happy to come and our project was judged in a positive light. The basic concept couldn't be simpler:

The Institute carries out scientific activities devoted to a current topic with the duration and scope of the activity to be determined by the program leader. The projects would be chosen by an international Advisory Board which would consist of a rotating panel of experts.

Everything was going according to plan. But then we had our first setback. Our original promised annual budget of 40 million shillings had been reduced to 10 million shillings and we had to completely rethink our concept. We thought that 10 million was also nothing to sneeze at; we just had to try to create a good working atmosphere at the very least if we couldn't be handing out riches. That's why I always aspired to treat our guests as great scientists and not as lowly supplicants. Unfortunately, this is something I had experienced before.

Naturally, the visitors got acquainted with the beauties of Vienna and excursions to the typical and picturesque Heurigen wine bars were popular. However, I refused to adopt the savings plan advocated by some financial gurus which called for paying our guests from the East less than those from the West on the grounds that they had less money at home anyway. That would have saved some money at first, but it would have poisoned the atmosphere, and once that has happened, you can't ever escape the bad reputation. The developments proved me right, and we never had to complain of a lack of visitors.

One of the most, if not the most, difficult problem turned out to be finding accommodations for the Institute. The University buildings were all already overbooked and there didn't seem to be a single spot free near the Institute, so that we had to cast our nets farther afield. But then Heide Narnhofer heard a rumor that culminated in a public announcement that a large, representative building was being offered close to the Institute.

On closer inspection, it turned out to be magnificently furnished and we already were picturing ourselves dwelling as exclusively as our colleagues who had settled in historical buildings. We euphorically invited our first guests, but in the next moment our bubble burst when the Ministry said that the building was too expensive. So we were left standing with our guests but without an Institute. But the Austrian proverb that "When God gives you a rabbit, He'll also give you grass" proved to be true.

Again Heide just happened to see an advertisement saying that there were two apartments vacant in the building where Schrödinger spent his last days.

It goes without saying that we grabbed on to this last life ring. The rooms weren't as spectacular, but they were inhabitable. The Institute even received a familial flair, as it contained a large kitchen that we left as it was. We immediately found out that a member of our advisory board, Pepo Marmo from Naples, turned out to be an excellent cook so that we could feast on Italian specialties every day for lunch.

Our finer psychological accomplishment, how we were able to convince the various committees that mathematics and physics would profit equally from the ESI, I'd like to pass over and come directly to our happy ending. The fact that the Institute could be opened in January 1993 is mainly thanks to the fortuitous constellation of having Erhard Busek as the Minister responsible. He held the same vision as I did. He was also keenly aware of Austria's responsibilities towards its neighbors to the East, and continues to devote all of his energies to this goal. The Ministerial Advisor responsible was Wolfgang Reiter, a trained physicist who knew intimately the strengths and weaknesses of Austrian science. He stood, and continues to stand after his retirement, ready to help the ESI in word and deed.

Does the ESI fulfill our expectations? Regarding the benefits for Austria, we can say yes. In the 15 years of its existence it has had an average of well over 100 visiting scientists a year. Furthermore, well over 100 scientific articles have been produced each year. The ESI continues to attract many scientists, although they only receive a little or even no money.

Regarding the benefits for our Eastern colleagues, we first must say that the picture has changed. The Eastern European countries were bought out intellectually by the West; most local researchers have left their native countries. Here in Austria, the feeling of solidarity for those remaining has gotten a bit lost. Some of the countries now have a fantastically wealthy upper class; the "new Russian" tourists in Austria spend on average 300 € a day; Austrians themselves only 100 €. You get the feeling that maybe it's time for this new wealthy class to start supporting their local scientists and to nurture and nourish science. In any case, it seems to me to be an untenable situation when our colleagues who are doing just as much science as we are, are only earning a tenth of what we do. But my only choice is to ignore this fact like everybody else; there's nothing else I personally can do. Moreover, our original motivations were also aimed at creating independent scientific schools in Eastern Europe and in this way to enrich the local culture with the new ideas that they would generate. This is certainly a major undertaking and it commits us to support these countries.

Entropy

To use an ice-skating metaphor, for Heide Narnhofer and me, the founding of the Schrödinger Institute was more of a compulsory exercise. Our "free skate" choice remains mathematical physics. She had industriously contributed to the four volumes of my course, which enabled her to gain an overview of where the ripest fruit was hanging in the garden of mathematical physics. Over the decades, the two of us have gone over all kinds of areas. I would like to mention at least one of these, which has entered the literature as C(onnes) N(arnhofer) T(hirring) entropy. CNT entropy is a variant of the usual entropy and tells you to what extent the system will continue to get mixed up by the dynamics.

But first, what is the "usual" entropy? Entropy is a peculiar hybrid; on the one hand it exists in thermodynamics as a quantity which expands through the amount of heat transferred. This is where the headache starts, as the question arises of how clearly is it defined which parts of energy are considered heat. When mosquitoes swarm, their energy is not called heat, but when molecules do, it is. Where do we draw the line?

Let's trust the experts and examine the other face of entropy. This is seen in information theory, where entropy is the lack of that information which is missing to the best possible knowledge. But this definition isn't much better and the headache's getting worse.

Who is informing whom? How is the information lost? Is someone withholding it? And can it be explained in more detail why this has anything to do with the previous definition? Heat is a form of energy and information is like a spirit in our mind without any objective reality.

Now the obvious objection that these quantities are of different dimensions can be easily overturned; they are the same except for one constant factor, the Boltzmann constant. With the appropriate units of temperature this constant can be 1. Our intuition tells us that the more atoms that are moving in a system, the more information will be needed for its complete description. However, CNT entropy remains rather heavy going, as it's about generalizing a classic term to quantum theory. Whoever thinks this would be too much for them can skip the following and continue after the box.

Without a doubt, entropy is fascinating. The fact that it can only increase was made into the second law of thermodynamics. Even though it is completely immaterial, it is the supreme ruler and dictates what may or may not happen. This it does in complete contempt of classical mechanics, in which for every motion the opposite motion is also allowed.

I hope to be able to verbally communicate our thought processes. There is so much, and so much that is nonsense, being written about entropy so that I can't assume that the reader has a clear understanding of what it is. Let me, therefore, first specify what is meant by the term entropy.

The term entropy was, as we've said, originally introduced in thermodynamics through heat and temperature, for our purposes it doesn't matter exactly how this was done. The only important thing is that it is not a material measure, or to put it another way, there's no place to buy a pound of entropy.

Nevertheless, it controls what happens. The second law of thermodynamics says that only those processes are allowed in which entropy increases. This law naturally consternated the atomists, who wanted to explain everything through the movement of atoms. However, the fundamental laws of mechanics are such that with every motion the opposite motion is also possible. How should a motion be forbidden macroscopically when it is the result of microscopic motions that are all individually reversible? Sure, we know that in particle physics the microscopic motions aren't completely reversible in time, but those are just tiny effects that we can ignore. Massive effects show that time reversal is not valid macroscopically. As Feynman so succinctly put it, "try to unscramble an egg".

New aspects came to light when Boltzmann delivered an easily understandable interpretation of entropy. He saw a large system that consisted of smaller subsystems, ultimately the atoms, which, however, we could neither see nor count. Proving atomism is so difficult because many different formations of atoms can look the same macroscopically: the macrostate can be realized by several different microstates. Boltzmann identified entropy as the logarithm of the number of microstates which lead to the same macrostate. Thermodynamic entropy is cumulative like an amount of heat, so the logarithm is used when two independent systems are being observed as a whole; the thermodynamic entropy is the sum of the two entropies. On the other hand, the number of states is multiplicative; every microstate of one can be combined with each microstate of the other.

So, entropy measures the information missing in the macrostate; the maximum amount of information would be when it was known which microstate had realized the macrostate. Entropy also provides a measure for the probability of when a macrostate will occur, assuming that all microstates are equally probable. To put in the image of rooms in the phase space we used previously, a microstate is a minimal room in quantum mechanics. All of them that look the same from a bird's eye view taken together build the macrostate. Their number is determined by entropy. Entropy as the logarithm of a natural number is either positive or zero; zero when the number is 1.

This is the case when the macrostate is a microstate, or when we have the maximum information about the system. To sum up: the greater the entropy, the more likely it is to find the macrostate, and therefore more information is won when a precise measure is taken of its contents.

A reasonable definition of entropy should solely concentrate on the basic properties of the state; the more random properties such as how the particles are numbered shouldn't matter. To put it in mathematical terms: structure preserving transformations called isomorphisms may not change the entropy.

This is where we run into the main difficulty. As the laws of nature are today still what they were yesterday, time evolution must be structure preserving and cannot increase entropy. There has been much talk about this embarrassing fact and the discussions fill many volumes. We will here simply take a look at a term from the Russian mathematician Andrey Kolmogorov which offered a way out of the dilemma. Kolmogorov considered partial systems of a large (infinite) system, and noticed at first that it also makes sense to speak about the unification of two subsystems. This unification is the smallest subsystem that includes both. Classical entropy is monotonous; it is greater for a larger system. Now, it's also possible to speak about the history of a subsystem, meaning the unification of all subsystems from which it developed, i.e. all of those which arise from the one at hand through the negative time evolution. In this kind of time story, the entropy of its history increases with time, since this is also getting bigger and bigger. In fact, the entropy of a history (from $t-\infty$) might be infinite, but in the change of this quantity with time, the infinity will be cancelled out. This also means that time will always educe something new and unknown from the subsystem.

According to Kolmogorov, "dynamic" entropy is the largest value that a subsystem can deliver per unit of time for the growth of entropy. "KS" entropy (S stands for Jakow Sinai) has many pleasant properties. Like entropy, it only depends on the structure of the system; it is the same for isomorphic systems. It even characterizes certain classes of systems. This means that for this class the converse is also true; if two systems have the same KS entropy, then they are isomorphic.

These were the findings that we discovered. The actual work began because Heide Narnhofer was invited to Leipzig so often. There she had been asked the obvious question about the quantum mechanic generalization of KS entropy; since, after all, nature obeys quantum mechanics. Now you'd think that this would be easy to answer, because the arguments given up to now are so general that they should have their counterpart in quantum mechanics. However, there's something to thwart our progress which haunts the literature as "entanglement" (and was given the awkward name "Verschränktheit" by Schrödinger). We encountered it with the van der Waals forces. These are the correlations of two systems which cannot be understood according to classical physics, but appear in quantum mechanics. (And also in reality, as Anton Zeilinger could measure with great precision). The consequence of this is that entropy does not grow monotonously with the size of the system; a partial system can have more entropy than the entire system. In particular, when you have the maximal information about the entire system you still don't have the maximal information about a partial system. As monotony is not a part of quantum mechanics, in it the KS entropy growth could under certain circumstances be a decrease; and that's not what we are looking for. This is the paradox that quantum statisticians had been watching with dread for years, and no one knew how to get out of it. However, for a special kind of quantum systems, in a certain sense the ones with an infinitely high temperature, the French math genius Alain Connes and his Norwegian colleague Erling Störmer found an expression of entropy which fulfills all desiderata. Heide and I were later able to show that in another extreme case, namely when the temperature is zero, that this entropy disappears, but these were only marginal phenomena; the entire area of finite temperature remained unchartered territory.

At that time, Heide and I were working on a contribution for writings dedicated to the memory of Jurko Glaser. He was my old friend from Zagreb, whom I got to know in Göttingen and who then ended up at CERN. My work with him was made a bit difficult because he was a chain smoker, and he also just began working when I was already thinking about going to bed. His desk was complete chaos, one layer of notes, a layer of books, then letters; in any case, there was no place to write. Our work together went like this: I scribbled down my half-baked ideas on a piece of paper and left it on his desk. The next day — miracle of miracles — I would find my ideas written down calligraphically, all of the mathematical tears had been mended and it was practically ready to be published as is.

After Jurko's death, they wanted to do something to honor him in Yugoslavia. I was naturally willing, but had my doubts if a special volume could be brought out in the foreseeable future. In any case, I wanted to produce something appropriate to his intellectual integrity. Heide and I were plowing through the fields looking at a quantity called "relative entropy". This measures how far apart the entropies of two states are. It is also commonly interpreted as the distance between two states. While we were looking at it from different sides, we stumbled on a quantity which shows a monotony just like what was needed by quantum dynamic entropy. Naturally, we included this monotony in our article and debated whether we should publish it separately in a more accessible journal as the definition of quantum dynamic entropy. Heide said that anyone can make a definition, but we didn't know enough about other properties.

In fact, Kolmogorov and Sinai's accomplishment does not consist of their having defined a quantity (KS entropy), but rather that they proved the universality of KS entropy. We were also in the situation where we had defined the entropy increase of a partial system, and for dynamic entropy this still had to be optimized for the partial system. This entropy was monotonous with respect to the partial system; therefore, for an increasing sequence of partial systems their entropy was also increasing and our quantum entropy should provide a universal upper limit. But we still wanted more: for every rising sequence of partial systems that is so large that it ultimately includes every partial system, the entropy should come arbitrarily close to our quantum entropy. It shouldn't be an exception of an artificial sequence, but something that is always present. In this aspect we were looking for a generalization of the equation from Kolmogorov and Sinai, because our definition for classic systems reduced itself to the KS definition. As the proof of the properties of the KS entropy is already quite difficult, we knew that its generalization would be even more problematic, and we put it on our list of problems to solve. The rest of the work we sent to Zagreb to be published in the Yugoslavian physics journal.

As this was not widely read elsewhere, we weren't worried that someone else would pick up where we left off and solve the problem before us. But here we were wrong. Somehow Alain Connes got a hold of it, and immediately comprehended the situation as he had worked out a special case with Störmer. He produced the missing proof and published it all in the Comte Rendues of the French Academy of Sciences (albeit with a mention of our work as well).

As soon as we saw this we invited him to give a lecture in Vienna and he told the whole story in a grand way. He suggested publishing the details of the theory in a joint article and naturally we agreed. Working together with such a top mathematician was an enriching experience. Not that he took care of the mathematical precision; his main concern was the conceptual organization. He said that our approach could be understood as the following: you must begin with the classic system by which you'd know what the KS entropy would be, but then you'd have to make a quantum correction. This is what's called a classic model, and then you look for the classic model which best approximates the quantum system given. The task remained to formulate the various suppositions and, in particular, the main result. We finally found reasonably straight proof and used it for our joint paper.

Du holde Kunst, a hymn to the art of music

Fig. 26: Prelude to my suite for the organ

Up to now I've been talking about how my professional career developed, but man does not live on bread alone. Through all of my travels, regardless of the direction I was heading in, I had a steady companion with me who brightened my spirit and to whom I owe many hours: music. The story of my musical education is quickly told. My parents sent me to piano lessons already as a child, but after two years I was considered to be difficult and without aptitude and sent to the piano teacher Maria Stubenrauch. Everything that I am able to do on the piano I learned from her. She also noticed that I was able to comprehend musical structures with ease, and she arranged for me to take theory lessons with Anton von Webern. At that time he was living in his villa in Mödling in a self-imposed exile, as the establishment rejected his music. He had to eke his meager living by giving private lessons. I then made the pilgrimage once a week to Mödling (right outside of Vienna).

Webern basically followed Schönberg's theory of harmony. This analytical method was a good fit for my way of thinking, and I was able to make progress quickly. I can still picture Webern standing in front of me, a thin slip of a man who was constantly fiddling with the fat cigars that would unfortunately cause his tragically early demise. After the war, he wanted to still smoke his cigar outdoors in the evening, despite the curfew imposed. When he lit the match, and American soldier saw the light, thought it was an attack and fired. This trivial misunderstanding extinguished the life of a genius.

After the war, I resumed my piano lessons with Maria Stubenrauch, and she sent me to two other theorists from Schönberg's circle, Joseph Polnauer and Erwin Ratz. I learned counterpoint from Polnauer and musical form from Ratz. They were in a similar situation like Webern; shunned by the establishment, they kept their heads above water by giving private lessons.

Later, I also studied for two semesters at the Music Academy with Josef Marx, but I learned more about craftsmanship from the Schönberg circle. I must hasten to add that they only taught classical music, i.e. not what you'd expect from the "second Viennese school". This was for the best, because serial music does not resonate within me and I would have just been producing dead constructs. But all of this still didn't bring me where I had actually wanted to go. When I was a teenager, I had a key experience in music when a friend asked me to help him actuate the stops of an organ. He was amazingly talented and much further along than I was. He could already play the Toccata and Fugue in D minor by Bach. I was already aflame by the mordent at the beginning. I was electrified (and still am today) and knew that I would have to be able to play that one day. However, as the pastor kicked us out I said to myself that I would one day have an organ of my own so that this would never happen again.

I would have to wait decades until I finally had enough time and money to practice a few hours on the organ each day. I had the good fortune to find a music professor, Leopold Marksteiner, who was able to deal with my by that time somewhat worn powers of comprehension. I have his talent in guiding his students to thank for the fact that I at least have a command over Bach's organ works.

For me, music is a miracle; it's impossible for me to describe it logically or put it into words. How is it possible that a series of air vibrations can burrow so deeply into the human psyche? How does it get into our heads? It can't be explained by the theory of evolution, for although the ability to yell loudly might be an advantage in natural selection, the ability to create a perfect fugue is not. Why is music such a universal language so that Chinese pianists feel the same thing that European ones do, as we can see in their gestures? Even animals understand this language. If I play the piano with the window open at our country home, I receive immediate and enthusiastic replies from all of the birds in the area.

For me, music is not only beautiful or ugly; rather many other associations come to play. It can be sophisticated or stupid, smooth or awkward, subservient or sassy. I ask myself where all of these ideas come from when it really is just some vibrations in the air? How did these connections enter our minds? In any case, music proves that there are things which cannot be explained by physical factors alone. Perhaps someone who's not musical and doesn't feel anything when listening to music would say that this is all delusion and fantasy which has somehow conned its way inside our nature through some genetic mistake. But our intuition tells us how wrong this scenario would be. It seems to me that music is the place where the human spirit most profusely bursts forth to the surface, using air as its material means of transportation. How else can we explain that soloists were able to perform more beautifully simply through Herbert von Karajan's power of suggestion than they could before? He was able to bring something forth which was created solely through mental interaction. Karajan did not have this power by himself, either. He could play the cembalo perfectly, but with others, he was able to go beyond himself.

At this, I will follow Ludwig Wittgenstein's advice: "what we cannot speak of we must pass over in silence." Rather than continue to speculate about music and its effects, I will do what I can which is make it resound. The reader may access the DVD via the link http://phaidra.univie.ac.at/o:1459. In order not to raise the wrong expectations, I must say that it is simply of house music. This serves personal edification when friends come to play music together. It is not music that's looking to explore new sounds, or provide a concert atmosphere flaunting acrobatic ability on the instrument. I have just expressed in notes that which pours forth from my soul.

The Computer

"What are you dressed up as, Papa?" My youngest son was looking at me as if he'd never seen me before. He thought that I was going to a costume party. I was wearing my cap and gown, but I wasn't going to a costume party, rather a different celebration; the 600th birthday of our University in 1965. To celebrate the event with due respect, all of the professors wore their celebratory cap and gowns and marched from the University to St. Stephen's cathedral where the grave of Rudolph the Donor lies. Although he was just in his 20's, he immediately set about seeing to it that his empire prosper. We did not inquire about which ways and means (e.g. fraud) he employed to achieve this, for us, it was only important that he founded the second German speaking University. This is why we wanted to honor him.

I will leave the question open as to whether our procession radiated dignity or was rather a picture inspiring pity. In any case, right beforehand a big fish got away. IBM was namely among the well-wishers and had offered us a noble birthday present: the largest and most modern supercomputer at the time. It was just that it would require a three-story building to accommodate it, which we were not able to provide. We were therefore not able to accept the gift.

The Technical School (today's Technical University) made the derisive comment that such a huge thing wouldn't have fit into our harmonious computer development, anyway. Why was it so big? It was rumored namely to have a huge 64Kb memory. None of us were computer experts, and we couldn't imagine what that meant, but it sounded wonderful. For those readers who avoid computers, a kilobyte (Kb) is what fits on a small page, a megabyte (Mb) the equivalent of a book, a gigabyte (Gb) an entire library. Today's mainframe computers operate with terabytes (=1,000 Gbs). I have saved my complete works on a 2Gb stick the size of a thumb, which could hold more data than 10,000 three-story IBM giants. (But naturally I'm no longer up-to-date. When Cornelia Faustmann, a young student, comes to me to work, she carries a small, stylish bag containing a portable hard drive with 120 Gb).

The processing speed has made enormous increases. We speak of "gigaflops", which are a billion processing steps per second. If you were to carry out a processing step each second, you'd need a billion seconds — something like 30 years — to be able to do what this tiny machine can do in a second. It's obvious that this ability opens up new possibilities in all areas of knowledge, and it would be foolish not to use them. This includes abstract mathematics, in which you start with a presumption and then try to formulate this as a theorem and then finally find the proof for it, where it's wise to first examine the extreme cases. Usually, that's where you see that you've bet on the wrong horse and then you don't have to waste any more time on it. When these controlled calculations take only seconds instead of months, you can naturally make progress much more quickly.

I had realized this early on, and when I was confronted with the work of an experimental physicist while sitting on a prize jury, I had the idea to get on board. The man had worked out a model for a thermal bath which was to give a system thermal balance, but the thermalizer only had one degree of freedom. The dominant theory at that time stated that a thermal bath must always be much larger than the system it is to thermalize. Without looking at the rest of the details, I immediately knew that the man was either delusional or exactly the right companion to come with me on my adventures and wanderings through statistical physics. After all, I wanted to test the limits. I thought that when he didn't shy away from ignoring the dogma of the establishment, then he also won't be frightened off by my wild ideas. This turned out to be the case, and for many years we did just what is usually forbidden in these circles.

Fig. 27: Harald Posch

We examined systems with negative specific heat, we did thermodynamics with systems with just a few particles when thermodynamics requires an infinite number of particles, we dared to explore systems in which, according to the theory, everything will become infinite, and we studied the many faces of instability. Harald Posch never hesitated or complained, he seemed to be having fun with it as well. I will select two of our adventures which are quite entertaining for general interest readers.

One of the things Harald Posch and I worked on together was examining the question to what extent the evolutionary biologists are correct when they say that the highly ordered biological structures have emerged purely by accident. These calculations can be understood even without special expertise, because the equations are so general that they can be applied to many other areas as well, particularly where there's money involved. Let's say there's a game with a certain number d of players (in science you'd talk about populations) and what develops is the fraction p_i, $i = 1,2 \ldots d$ which the player i possesses from the circulating capital. (We'll choose a currency in which the capital is 1). The course of events is determined by a matrix m_{ik}, which shows which percentage of his wealth player i *transfers to player k in a certain amount of time. We will use only the credits, instead of debits; we'll have a flat tax. This means that the same percentage is taken out of every credit. This percentage is calculated so that the amount of all of the capital remains 1, so that every sum total of* $p_i = 1$ *(written* $\Sigma_i p_i = 1$*) is always valid.*

In these financial terms, order is when one player possesses all of the money, so one $p_i = 1$*, the others 0. Chaos is when everybody has about the same amount* $p_i = 1/d$ *for every i.*

Entropy is used to measure the order: entropy = $1 - \Sigma_i p_i^2$*. It is zero when one player has everything and 1/d when everyone has the same amount, otherwise it's somewhere in between these two values. But where is the random coincidence? What this is exactly I also don't actually know, but I'll assume that the matrix* m_{ik} *has been created by the conditions that dominate at the time and is therefore random. What does this statement tell us? The values of* m_{ik} *are positive and, let's say, limited by 1. Let's split the interval between 0 and 1 into many parts. These are the values that we can enter into the* d^2 *square of the* m_{ik} *successively, and solve the evolutionary equation for everyone on the computer.*

It turns out that after a long time the p_i *will typically aim for a limit which is independent of where it started. This means that it survives the best possible constellation. This can also be more complicated. It doesn't have to be that there's one* $p = 1$ *and that the others are 0. The p give us the value for entropy, so that when we have been working hard and have solved the evolutionary equations, we'll get an immense number of values for entropy, with the dynamic being chosen randomly by the best possible a priori principle. In case these entropy values should all be close to the maximal value, then we say that the system is mixing. Providence was not able to triumph over chaos. If they hover at 0, then Providence has selected one lucky winner, the most capable has survived.*

Which of these alternatives actually happens? Contrary to the evolutionists' expectations, the system is mixing, when all m_{ik} *are allowed, pure chance levels everything out. If, however, there are some limits placed on the* m_{ik} *and only the area* $i > k$ *or* $i < k$ *is allowed then the tide turns, and only one prevails; it's completely random which one. This is also easy to understand. The relation* $i < k$ *creates a hierarchy amongst the players: the greater the index i the higher the player is in the hierarchy. The financial transactions are such that the player who receives has a higher rank than the player who gives, and it's never the other way around.*

Under these asocial circumstances, it's no wonder that the player on top will possess everything after a certain period of time. The question is what do we need to start a revolution, or how much do the higher-ups need to pay to the others so that their wealth is balanced, i.e. so the entropy grows. The answer is; surprisingly little. When the m_{ik} *for* $k < i$ *make out only 2% of those with* $i < k$*, the reversal already starts to take place, and an equal distribution will result. So it's fair to say that only when order is forced using the most brutal methods does it emerge; otherwise chaos reigns. From a population theory point of view it should first be said that Darwin was right in general. The "fittest" survived. It's just that the fittest is generally not a species, but a constellation of species living together symbiotically. Only under special conditions is only one the fittest.*

This kind of general overview is only possible thanks to today's high-speed computers. They can solve the d equations for the p_i *in a fraction of a second, so that the million* m_{ik} *have been tried and tested in just a few minutes.*

The computer was also able to contribute new aspects to the oldest problem of physics, planetary motion. We can now calculate orbits in a fraction of a second when it used to take us years to do so, and change the masses, distances or eccentricities at our discretion. In this way, you can find out fast what's important. The story is so amusing that I'd like to tell it from the very beginning.

The planets were already treated with suspicion in the Middle Ages due to the way they slipped around the fixed stars, and because of the lack of other suitable personnel, angels were given the task of making sure there were no traffic accidents. Even Newton was still under the assumption that God had to adjust the orbits from time to time to avoid a catastrophe. This was challenged by Laplace. When Napoleon asked him where God was in his system he simply replied, "I don't need this hypothesis."

Today we have to be amazed by Newton's visionary power, even though it was long considered to be residual superstition. Laplace was certainly correct for the time periods of thousands of years, which is how people thought at that time. The predictive power of Newton's equations even stretch over millions of years. However, today we know that the solar system is 4.5 billion years old, and in order to predict things, we would need a completely unrealistic level of precision for the current condition. There was no way for Newton to know all this, but somehow he suspected it. However, Laplace didn't have a real proof for stability either, and the pattern of proofs and their criticisms didn't seem to want to end.

Finally it got to be too much for the Swedish King Oscar II. and he offered a sizable reward for anyone who could provide conclusive proof of the stability of the solar system. Towards the end of the 19th century this attracted all the great mathematicians of the day, and the jury finally selected Henry Poincare, who was the leading French intellectual at the time. The Swedes felt so honored that they founded a new scientific journal, which is renowned up to today. Poincare's work was to grace the first issue, but then came the upset. Poincare found a spot in his proof that was not airtight; his proof wasn't conclusive or even proof. He wanted to withdraw the article, but it was too late, he had to have the entire first edition destroyed. This cost him more money than he had won in the prize.

Still, it wasn't entirely a loss for Poincare. He did develop the right way to think about this problem, so that his name is still present in this branch of science. For the system of the sun plus two planets, the Swedish mathematician Sundman was able to prove what Poincare had tried to a few years later. More precisely, he showed that the equations of motion for all periods of time predict what will happen. What stands in its way are collisions which make it mathematically impossible to predict what happens next due to the singularity of the potential.

Sundman had argued that collisions in which all three are in the same place could not occur when the total angular momentum is not zero. For a collision of two things he introduced a new time variable which disperses the accelerating events. This seemed to save stability, but then something else happened. Two American mathematicians were able to construct an orbit with four particles in which one body slams the others so far apart that after a finite period of time they have already reached infinity. They didn't say what the particle does when it gets there, but in any case, stability is refuted, time doesn't go any further. Some diehards didn't want to give up, and made the objection that two-body collisions occurred, which isn't allowed.

But then a Chinese mathematician found an orbit with five particles in which the same thing happens with only near-misses, so there is no stability.

Over time people realized that not only the aspect dealing with stability, but the whole question was wrong. In the first place, the Twelfth of Never $t = \infty$ is completely immaterial for us; our solar system will only exist for 5 billion more years. In the second place, the question of whether a motion can be mathematically continued doesn't matter to us; we need a stable climate, meaning we need to have a constant distance to the sun over this time. This can only be guaranteed for the next ten million years. The solar system runs like clockwork, there will be no catastrophe because it is a top product of Darwinian selection. All orbits are almost circular, they don't cross over each other (Pluto was disqualified). The large planets are far out and stabilize the orbital plane with the angular momentum, and they catch all of the cosmic debris for us. New verses are constantly being composed for their song of praise.

The first four billion years must have been pretty rough going in our solar system. The asteroid belt bears witness to hard collisions. Our good old moon is a part of the Earth which was ripped out by a wild neighbor. A typical catastrophe is when two planets get so close to each other and one gets so much momentum that it gets catapulted out of the solar system. Countless victims in this struggle of natural selection have already gathered way out there. Many more have certainly been disposed of without a trace in the sun. However, the specifics of all of this are based on speculation. Beyond the ten million years, no guarantees about the details can be made; they are too dependent on the original and/or the final conditions. This entire dynamic is at first just a fantasy, but can be brought to life on the computer when the planetary orbits are calculated with different masses for the planets, and their distance to the sun, their eccentricity and their energy are varied. Then our solar system does not appear as a harmless clock, but as an arena for the fight of the fittest planet to survive. This was the kind of game we watched on the screen with great emotion.

The Cardinal König Encounters

"This is just like in Lessing's 'Nathan the Wise'", my old friend Walter Kohn said, "Here we three are sitting like his wise men, each of us has a different religion and yet we still can communicate with each other so well." Walter Kohn is Jewish, I'm Protestant and the third was Cardinal König, who had been so kind as to invite us to lunch. Walter had once remarked to me that one advantage of the Nobel Prize was that it opened all of the doors, and then you could meet people like Cardinal König. I replied that he should let me know the next time he came to Vienna, as Cardinal König would certainly be pleased to meet him. This is what happened and after a warm welcome, the Cardinal, in his unobtrusive way, knew how to immediately direct the conversation to the things that matter.

We were therefore soon discussing theodicy. Walter said, "Since the Shoah, I can no longer believe in an omnipotent God. How could He have allowed that to happen?" The Cardinal, who always took into account what other people thought, turned to me and asked, "How do you see this, Professor?" At the time I had just finished reading a lot about it, and had the impression that a lot was hedging going on. Some, like Teilhard de Chardin in the addendum to his book, admit that they really can't understand it, and some are satisfied by uttering some random statement. I thought that the least objectionable thing to do would be to offer a quote:

Me: "The best minds have been racking their brains and scientific knowledge doesn't help here, either. Perhaps the correspondence of two renowned mathematicians about this topic could shed an interesting light on the subject. One of them, Vietoris, was an extremely pious man. He didn't only go to mass every morning at 6 a.m., but legend has it that his wife and even all six of his daughters had to go with him. The other, Gröbner, was a militant atheist who was constantly arguing that the theological departments should be shut down at the universities as taxpayers' money shouldn't be spent on this unproven stuff. Despite the differences in opinion, they never got out of line in their letters, the gist of which was:

Gröbner: How can you believe in a benevolent God when there are such horrible things just spilling out in the world?

Vietoris: You say that the world is so bad, but maybe it's just bad here, maybe it's more perfect somewhere else?

Gröbner: I couldn't care less about what it's like in other solar systems, the only thing that matters to me is what's happening on Earth, and here it's really bad.

Vietoris realized that Gröbner couldn't be so easily placated and that he would have to think of something better. So he continued with: God gave us freedom and lets us do as we wish even when we do something evil. Our freedom means that there has to be evil in the world.

At this point ***Cardinal König*** cut in and said that, "these are positions that you hear all the time, but what do you personally think?"

Me: Your logic, Walter, can't be undermined. This police-God who looks at every detail taking place on the Earth and ensures that nothing bad happens anywhere; this obviously doesn't exist. This can't exist, because there are also different opinions about what good and evil are. This was obvious in the religious wars in which both sides had their weapons blessed so the just cause would win. I think here we have to admit that we don't understand and listen to what Jesus said, "take what is yours and go".

Kohn: "I can't accept that, that's too defeatist. We must actively fight against evil!"

And this is what he did the next day, when he flew back to America to spur on his anti-Bush campaign (unfortunately without success).

With his open, charismatic personality, Cardinal König brought a new spirit into the Catholic church. When someone asked him once if he wouldn't like to become Pope, he said that he wasn't made out of the stuff of Popes. I saw it the other way around, that unfortunately the Popes had not been made out of his kind of stuff.

I personally shouldn't complain about Popes. I was chosen to become a member of the Papal Academy of Sciences, and we are always treated very well in the Vatican. The Academy is in a palace located in the middle of the Vatican Gardens and we are able to enjoy all their beauty during the breaks in our meetings. We stay in a hotel inside the Vatican, directly across from a back door in St. Peter's. Early in the morning we can sneak in before the cathedral is overrun with tourists and spend time alone with Michelangelo's Pieta. However, the Academy is not just a spa for its members; it also serves a good cause. For example, during the nuclear arms race, John Paul II sent a delegation of the Academy with a personal message to the presidents of the USA and the USSR, demanding an end to proliferation and nuclear disarmament. Perhaps the Academy carries such weight because its members are selected without regard to their religion, nationality or social class; only their scientific achievements count, and these are not subject to any dogmatic framework. In any case, soon thereafter the nuclear disarmament treaty was signed into effect.

When the Templeton Foundation asked me if I would be willing to organize discussions between theologians and scientists, I saw an opportunity to show Cardinal König what I had been referring to. I agreed and called this the Cardinal König Encounters. Naturally, the project was met with a mixed reception; some people thought I was having a kind of fit of religious fervor caused by the onset of old age.

I still considered the positive aspects as outweighing the negative. There were participants representing all different religions — atheism is, for me, a religion as well — the young and old alike attended, and for many it was liberating to be able to talk about everything in a relaxed atmosphere. Of course, I realized that one side wouldn't be able to solve the problems of the other, but at least misunderstandings could be cleared away. We as people are all in the same existential situation. We arrived in a world which is on the one hand ruled by highly cerebral laws, but, on the other, a lot of things seem to happen purely by chance. In this way, highly complex structures were created from the original chaos of the Big Bang, namely people. According to modern scientific understanding, this happened due to a long chain of fortunate conditions. This mystery is so challenging that it shouldn't be dismissed with an agnostic shrug of the shoulders. Regardless of whatever image of God we've been handed down or developed ourselves, we must be thankful for having been created and we must fulfill our potential of what we were made to do.

Many scientists complain that the Creation seems meaningless to them. As the word "meaningful" isn't first defined by science, they can't decide if our existence is meaningful or meaningless. For me, only an inner voice can tell me what being meaningful means and I have to decide if my existence is or not. Whatever images and perceptions we have wrested from the Creator, we owe humility and recognition to the fact that we arrived in such a wonderful world.

What has Scientific Research Achieved?

I began with the idea of how much our environment has changed since I was a child. However, our understanding of the makeup of the material world has undergone even more drastic changes. Consciously or not, a materialistic-deterministic worldview was at the back of people's minds at the beginning of the 20th century. According to this, the world is made up of atoms. Atoms are tiny indestructible bodies, between which there are forces which dominate; a weak attraction over greater distances and a strong repulsion over small distances. Everything moves under these forces according to the laws of Newtonian mechanics. When you know the location and the speed of all atoms at the beginning, then the future can be completely determined.

Atoms are everything in this worldview. They were broken down into electrons and atomic nuclei at some point, but otherwise there is nothing else. Some things, like light, could not fit into this picture, and therefore a mechanical model was desperately being sought after for electromagnetic properties. People thought that mechanics had the final word and a phenomenon could only be truly understood when there was a mechanical explanation for it.

Now, this simple mechanical image enjoyed many great successes; for example, it can provide a plausible explanation for a lot of things in thermodynamics, even if it can't prove them. That's why it was so convenient to adhere to this model and just keep brushing smaller incongruities aside to be dealt with later.

This picture was re-examined in the 20th century. Not only were the smaller wrinkles ironed out, but the entire picture was completely abandoned because it was ontologically wrong. This scholarly term is used to imply that the being of things was not being reflected correctly.

I'd like to illustrate the distinction between wrong and what is again wrong using the transition from a geocentric to a heliocentric world system. In a geocentric system, the Earth is at rest and the stars are moving around it, and in a heliocentric system everything revolves around the sun. Today there's no doubt that the latter is correct and the geocentric is wrong, even though it describes everyday life perfectly. This judgment can be relativized, and the objection can be made that according to Einstein's general theory of relativity, every reference system is allowed. In our everyday language we talk about things as if the Earth were standing still, so maybe both systems really are equally good.

However, this is not the case: the geocentric system is ontologically wrong. We don't even have to start with the question of who determines what is moving, but rather the picture of the universe that this system imparts is a complete distortion of reality. It provides the idea that the Earth is resting at the center and dominates the entire universe, all around it the stars are revolving as marginal phenomena and that's it. This picture is utterly wrong. The Earth, measured against the vast expanse of the universe, is just a speck. Everything that we can see without additional equipment is just a small sliver of a much greater cosmos. The question of who is now "really" moving can't be answered; actually, no one is at rest.

The situation is similar with the mechanical model of the world. The material particles have disappeared from the first row and been replaced with quantum fields. These have the state of the highest symmetry, the vacuum, in which the particles exist only virtually as a possibility. The real particles are only local excitations of this state, really just like a bit of kicked up dust. They have also surrendered their individuality, in so-called Bose–Einstein condensation the particles join together in a collective in which their number is not clearly defined. Today's standard model of matter replaces the atomic nuclei from the earlier picture of matter with quark fields; the electrons with lepton fields; the forces between them with gauge fields and Newtonian mechanics with field equations. A standardization was achieved; force and matter are both quantum fields that simply obey different statistics. Materialism was dismissed insofar as the term matter no longer plays a special role; the new protagonists are energy, symmetry and structure.

You could think that now the stumbling block between a scientific and religious worldview had been cleared away, but this is not the case. The stumbling block consists of the small follow-up "and that's all."

If you now say that the universe is nothing but a collection of atoms or of quantum fields, the theological side will object that there's still something missing. They are the two words "nothing but" which can be both true and completely miss the point. When you say that all of mathematics can be stored on a computer and as a computer only knows series of 0 and 1 then mathematics is just a series of 0 and 1, well, that's true, but still nonsense.

What's missing isn't another field or other kind of agent, but the insight as to all that can be developed out of the theory. The Schrödinger equations rule atom theory, for instance. But if you didn't know anything about life, you would never have been able to deduce what complex compounds atoms can build, or the wide variety of living things these continue to develop.

Despite the wealth of modern knowledge, it may be that this is just a small part of a much more comprehensive reality than is accessible with our methods. There are no equivalents for anything spiritual in our theories of physics; the spiritual still await their spark of inspiration. Neurologists like to say that there is nothing spiritual, that in reality it's all just a storm of electric impulses in the neurons. Again the "it's all just" is an ideological addition — a translation from Hottentott to Zulu — without which the statement would be incontestable. When the Theory of Everything is found and we reach the deepest layer of events, even then research won't stop. We remain standing where Newton was when he said: "I was like a boy playing on the seashore, and diverting myself now and then finding a smoother pebble or a prettier shell than ordinary, whilst the great ocean of truth lay all undiscovered before me."

Picture Credits

Index of Names

Subject Index